Shaky Foundations

Studies in Modern Science, Technology, and the Environment

Edited by Mark A. Largent

The increasing importance of science over the past 150 years—and with it the increasing social, political, and economic authority vested in scientists and engineers—established both scientific research and technological innovations as vital components of modern culture. Studies in Modern Science, Technology, and the Environment is a collection of books that focuses on humanistic and social science inquiries into the social and political implications of science and technology and their impacts on communities, environments, and cultural movements worldwide.

Matthew N. Eisler, *Overpotential: Fuel Cells, Futurism, and the Making of a Power Panacea*

Mark R. Finlay, *Growing American Rubber: Strategic Plants and the Politics of National Security*

Jill A. Fisher, *Gender and the Science of Difference: Cultural Politics of Contemporary Science and Medicine*

Finn Arne Jørgensen, *More than a Hole in the Wall: The Story of What We Do With Our Bottles and Cans*

Gordon Patterson, *The Mosquito Crusades: A History of the American Anti-Mosquito Movement from the Reed Commission to the First Earth Day*

Thomas Robertson, *The Malthusian Moment: Global Population Growth and the Birth of American Environmentalism*

Mark Solovey, *Shaky Foundations: The Politics–Patronage–Social Science Nexus in Cold War America*

Jeremy Vetter, *Knowing Global Environments: New Historical Perspectives on the Field Sciences*

Shaky Foundations

The Politics–Patronage–Social Science Nexus in Cold War America

MARK SOLOVEY

RUTGERS UNIVERSITY PRESS
NEW BRUNSWICK, NEW JERSEY, AND LONDON

First paperback edition, 2015

Library of Congress Cataloging-in-Publication Data

Solovey, Mark, 1964–

Shaky foundations : the politics–patronage–social science nexus in Cold War America / Mark Solovey.

p. cm.

Includes bibliographical references and index.

ISBN 978-0-8135-5465-5 (hbk. : alk. paper) — ISBN 978-0-8135-5466-2 (e-book)

1. Social sciences—Research—United States—History—20th century. 2. Endowment of research—Political aspects—United States—History—20th century. 3. Cold War—Social aspects. 4. World politics—1945–1989. I. Title.

H62.5.U5S65 2013

300.72'073—dc23

2012012098

A British Cataloging-in-Publication record for this book is available from the British Library.

Visit our website: http://rutgerspress.rutgers.edu

Manufactured in the United States of America

For Marga, for my mom, and for my dad

CONTENTS

ACKNOWLEDGMENTS

It is a pleasure to thank the following individuals, who offered cheerful encouragement and helpful feedback: Toby Appel, Roger Backhouse, Michael Bernstein, Allan Brandt, Howard Brick, Michael Bycroft, Jamie Cohen-Cole, Hamilton Cravens, Colleen Dunlavy, David Engerman, Matthew Farish, Philippe Fontaine, Daniel Geary, Christopher Green, Emily Hauptman, Ellen Herman, Hunter Heyck, Sarah Igo, Juan Ilerbaig, Joel Isaac, Andrew Jewett, Edward Jones-Imhotep, David Kaiser, Paul Kingston, Daniel Kleinman, James Kloppenberg, Nikolai Krementsov, Rebecca Lowen, George Mazuzan, Neil McLaughlin, Jal Mehta, Michael Pettit, Wade Pickren, Jefferson Pooley, Theodore Porter, Julie Reuben, Joy Rohde, Marc Rothenberg, Alexander Rutherford, Laura Stark, Marga Vicedo, Jessica Wang, Nadine Weidman, and Andrew Winston. For providing amusement and perspective, I am grateful to the marvelous HAT.

I am also thankful to a number of institutions, organizations, and academic units that provided funding and other resources that facilitated my research and writing: the U.S. National Science Foundation; the Institute for the History and Philosophy of Science and Technology, the Centre for the Study of the United States, Victoria College, and Victory University, all at the University of Toronto; the Charles Warren Center, the History of Science Department, and the History Department, all at Harvard University; the Agence Nationale de la Recherche in France, research programme in cross-disciplinary research ventures in postwar American social science; the Max Plank Center for the History of Science in Berlin; Arizona State University West; and the Science and Technology Studies Program at MIT.

For helping me to locate and use research materials needed for this study, I am indebted to the professional staff at the following libraries and archives: Columbia University Center for Oral History, Ford Foundation Archives, Harvard University Archives, Library of Congress, National Academy of Sciences Archives, National Archives, National Science Foundation Library, Rockefeller Archive Center, Smithsonian Institution Archives, and University of Chicago Library and Special Collections Research Center. Wiley, the publisher, deserves my gratitude for granting permission to reuse some material from my article "Riding Natural

Scientists' Coattails onto the Endless Frontier: The SSRC and the Quest for Scientific Legitimacy," *Journal of the History of the Behavioral Sciences* 40 (2004), 393–422.

Last, I want to recognize the team at Rutgers University Press who turned my manuscript into this book. Special thanks to Peter Mickulas, my editor, who gave me excellent advice at many points along the way and did a superb job shepherding this work through the publication process.

Shaky Foundations

INTRODUCTION

Social Scientists and Their Patrons in a Remarkable Era

From the mid-1940s to the mid-1960s, the social and psychological sciences in the United States experienced dramatic growth. In 1947, the American Psychological Association, the major professional society for psychologists at the time, had 4,661 members. Within a decade the APA had 15,545 members, and by 1967 25,800 members. The American Sociological Association grew fivefold during this time period, from 2,218 to 11,000 members.[1] Fortified by wider developments in American higher education and science, vigorous expansion also marked the trajectories of university departments, research centers, graduate programs, academic journals, and scholarly publications in the social sciences. Numerous new areas of social research flourished as well, including many interdisciplinary fields, from systems analysis to the behavioral sciences.[2]

Professional and academic growth contributed to greater visibility and influence of the social sciences in national affairs. During the first two post–World War Two decades, many scholars penned studies that gained renown beyond the ivory tower, as did a few journalists who drew heavily on their work. Among these successful authors are Stuart Chase, Erik Erikson, Milton Friedman, John Kenneth Galbraith, Michael Harrington, Alfred Kinsey, Margaret Mead, C. Wright Mills, Gunnar Myrdal, Vance Packard, David Riesman, Walt Rostow, B. F. Skinner, Benjamin Spock, and William Whyte. Meanwhile, the work of new federal agencies, from the Council of Economic Advisers to the Central Intelligence Agency, suggested that social science knowledge and expertise had a central role in achieving national goals. In the early 1960s, none other than the new and youthful American president expressed strong enthusiasm for working closely with the nation's best and brightest. As President Kennedy saw it, action-oriented intellectuals had great contributions to make on the home front and in the global battle against the communist menace. During the Kennedy and Johnson administrations, economists, sociologists, and scholars from nearby disciplines contributed to bold domestic and foreign policy initiatives, from the War on Poverty to the war in Vietnam.

As a rich body of historical literature in the last couple of decades has shown, the rapid growth of scholarly social research and the many policy contributions of the social sciences were intimately tied to their relations with extra-university funding sources. Federal patrons, including military, propaganda, and intelligence agencies, civilian science agencies such as the National Science Foundation and the National Institute of Mental Health, and the large private foundations created with the wealth of the Carnegie, Rockefeller, and Ford fortunes, all took a strong interest in the social sciences. Patron-funded work, in turn, promoted influential ways of understanding and managing high-profile social and political issues. In recent years, as the literature on the social sciences in Cold War America has acquired increasing depth, interest in "following the money" has emerged as a central theme. Contributions to this literature have come from a variety of perspectives, most notably from the history of science, the history of particular disciplines such as sociology or economics, and intellectual history, but also from some fields that one might not initially expect, such as diplomatic history.[3]

Yet only a small part of this literature has taken the development of funding sources and their engagements with the social sciences as their principal focus. Discipline-oriented histories, which have always represented most of the historical literature on the social sciences, have typically concentrated on the major schools of thought, leading departments, and prominent scholars within a single discipline. These studies sometimes note that successful scholars and research programs received funding, which, in turn, contributed to their status within the discipline. But with few exceptions, these studies have left aside careful attention to patrons themselves.[4] Even influential accounts that discuss the impact of funding often stop short of examining the development of patron interests, policies, and programs more directly.[5] Neither have authors paid much attention to social scientists' efforts to influence the ways funding sources defined and supported their work. Yet the leaders and governing boards of private and public funding sources recognized that they needed guidance from scholars who could provide insights regarding the historical development of the social sciences, their present status, and promising opportunities for their future development. As a result, a select group of scholars became research consultants, policy advisers, and program managers, in the process producing many documents that helped to establish the relevance of the social sciences to patron concerns.[6]

How patrons should be included in histories of the social sciences that tell big-picture stories has remained unclear as well. Existing studies that discuss the influence of patrons have nearly always focused on a particular discipline or specific field of research. As a result, the broader significance of patrons and their engagements with the social science enterprise needs much more explicit attention. Moreover, synthetic histories of the social sciences have hardly paid any attention to the evolution of the national funding landscape. This lack of attention surely reflects the long-standing tendency to use the history of ideas as the basis for constructing big-picture stories.[7]

This book advances our understanding of patrons and their wider importance for the history of American social science by examining a small group of new funding sources during the early Cold War decades: specifically, the National Science Foundation (NSF), the military, and the Ford Foundation, especially its Behavioral Sciences Program (BSP). I analyze the emergence and influence of these patrons, including the development of their policies and programs for supporting the social sciences. In doing so, I pay close attention to the roles of scholars who worked closely with these patrons and thus had special opportunities to influence, promote, and assess their efforts to promote progress in this area. As this study will show, the military, the Ford Foundation, and the NSF acquired special importance by providing social scientists and other interested parties, including some influential politicians and natural scientists, with new opportunities to work out the nature and uses of the social sciences in the anxious nuclear age.

The Argument: One System, Two Commitments, Five Challenges

My central argument consists of three related claims. First, the patrons under consideration here became crucial components in an enlarged and transformed extra-university social science funding landscape. During the era between the two world wars, the most prominent extra-university patrons had been the large private foundations, especially those created with the great Rockefeller and Carnegie fortunes, and some federal agencies, including the Department of Agriculture.[8] In the early 1940s various agencies involved in the war effort turned to the social sciences for help. But at the war's end, the future of private and public funding for the social sciences was rather uncertain. Thus in the late 1940s and early 1950s, at a time of pervasive debate about the nature and purposes of the social sciences within philanthropic circles and the national science policy arenas, new patrons had the chance to establish their prominence quickly. Though strictly speaking the military did not qualify as a new patron in the postwar years, the extent of the military's involvement with the social and psychological sciences remained uncertain at the war's end as well.

As the military, the Ford Foundation, and the NSF established policies and programs to support the social sciences, they did so by defining their responsibilities and activities in relationship to one another. This type of informal coordination makes it useful to consider these patrons in terms of a single, albeit loosely integrated system. Though this system did not develop through centralized control, in many important respects these patrons developed their policies and programs in relationship to one another. Moreover, informal coordination supported striking commonalities in their efforts to advance the scientific and practical value of the social sciences, which leads to the next main claim in my argument.

Second, these patrons, together with the social scientists who worked most closely with them, embraced a strategy that rested on two key commitments, to scientism and to social engineering. The first commitment involved accepting, in a broad sense, a unity-of-science viewpoint, which assumed that the social sciences lagged behind the more mature natural sciences and which posited that the former should follow in the footsteps of the latter. Often this viewpoint meant the social sciences needed to rid themselves of their involvement with a wide array of humanistic forms of inquiry, including "soft" qualitative, philosophical, historical, and normative forms of analysis. Just as importantly, social scientists had to establish a clear distinction between scientific social inquiry and other value-laden spheres of social action, such as politics, social reform, and ideology, and especially Marxist or socialist perspectives. More positively, this viewpoint implied that the path to scientific credibility and progress lay in the pursuit of more rigorous, systematic, and quantitative investigations that promised to yield accurate predictions about what individuals, social groups, and social systems, including economic and political systems, would do under stated conditions.

The other key commitment, concerning the social sciences' practical value, indicated that this work would contribute to the national welfare and human betterment more generally through social engineering applications. This commitment often rested on an instrumental viewpoint, which regarded social science knowledge, techniques of analysis, and expertise as apolitical, nonideological, and value free. A very common idea associated with this position suggested that basic or pure scientific inquiry, whether in the social or the natural sciences, produced value-neutral knowledge of a fundamental sort. Such knowledge, in turn, provided the basis for realizing desired practical goals in a couple of ways, depending on the specific domains of investigation. Certain lines of investigation sought to place the processes of decision making on a rational basis. Other lines promised to facilitate control over individuals, social groups, and social systems. Both manners of realizing social sciences' practical value rested on a technocratic outlook, as their proponents generally assumed that leaders and managers in various sectors of society, especially in government, comprised the most relevant audiences for social science knowledge. Thus these leaders, rather than the American public more generally, would be ultimately responsible for using this scientific knowledge wisely.

A number of historical accounts have already noted the prominence of scientistic and social engineering commitments in the development of individual disciplines, subdisciplinary fields, and interdisciplinary areas of study during the Cold War years.[9] But such accounts can at most offer only informed speculation about why such commitments became so ubiquitous throughout the social and psychological sciences. By examining how those commitments became so prominent in the new funding landscape, this study will help us to understand how and why they acquired a pervasive presence throughout these sciences. Excellent historical studies have already demonstrated that such commitments had strong support prior to the Cold War era. However, I propose that we should not assume a straightforward path in their development from the prewar to wartime to Cold War contexts. Later in this introduction I will return to this question about how to narrate the longer story of scientistic and social engineering commitments in twentieth-century America and explain why I think the Cold War funding landscape deserves a prominent place in this story.

The third main claim advanced in this book concerns an array of obstacles and criticisms that new patrons and their scholarly collaborators faced as they sought to promote the social sciences based on a scientistic strategy. Here, I identify these obstacles and criticisms very briefly, in the form of five challenges:

1. While conservative interests put pressure on social science patrons and scholars to distinguish their work from left-leaning positions, conservative criticisms also pointed in two contradictory directions, on the one hand by suggesting that social science could and should become more like the natural sciences and on the other hand by proposing that allegedly value-neutral, apolitical work distorted social inquiry, undermined the nation's religious, political, and moral values, and placed established institutions in danger.
2. Some liberal critics also argued that trying to imitate the natural sciences had disturbing consequences, but, contrary to conservative critics of scientism, they suggested the social sciences should be directly engaged in critiquing an unjust status quo and promoting progressive reforms.
3. Regardless of one's particular political views, many scholars recognized that the influence of patrons threatened the independence and quality of scholarship.
4. The new patronage system encouraged social scientists to emulate natural scientists, but within the federal science establishment social scientists were never able to overcome their second-class scientific status and associated difficulties produced by the reigning scientific pecking order.

5. Despite the widespread claim that social science could and should steer clear of value-laden work, certain lines of research supported by the Ford Foundation and the military made it increasingly difficult to draw a sharp boundary that would clearly separate scientific social inquiry from ideology, politics, or reform.

When considered together, these five challenges indicate that the new patronage system had noteworthy vulnerabilities and weaknesses. Recognizing these challenges also helps explain why patrons, scholars, and their scientistic and social engineering commitments all came under substantial scholarly and political scrutiny starting in the mid-1960s. Given the wider historical implications of these challenges and their importance in the present study, I will say more about each of them later in this introduction.

The following chapters present detailed case studies to support my argument and its central claims. Chapter 1 examines the place of the social sciences in a pivotal national science policy debate during the mid- to late 1940s that concerned competing legislative proposals for a new national science agency and that eventually led to the establishment of the NSF. In this debate, some influential and mainly conservative members of the nation's natural science, medical, engineering, and political communities supported legislation based on a set of science policy principles recommended by Vannevar Bush, one of the nation's leading scientific statesman, in his famous 1945 science policy report, *Science—The Endless Frontier.* This group supporting Bush's views also presented many objections to including the social sciences within the purview of the proposed national agency, in the process revealing a deep well of suspicions that suggested these disciplines lacked a solid scientific foundation and, even more damning, that indicated they had more in common with leftist ideology, social reform, and politics than with objective scientific inquiry. Meanwhile, prominent social scientists from the Social Science Research Council (SSRC) developed a strategy for gaining inclusion that rested on a unity-of-science viewpoint. Led by the eminent economist Wesley Mitchell, the SSRC contingent claimed, contrary to their conservative critics, that modern social science inquiry had an admirably scientific character and shared a great deal in common with natural-science inquiry. They explicitly dismissed alternative viewpoints that held scholarly research should be directly involved with moral, social, and politics goals. In addition, they argued that social science had valuable social engineering applications based on the development of apolitical and ideologically neutral knowledge. Yet, in the end, their critics managed to place the social sciences on the sidelines of the ongoing national science policy debate. This outcome, moreover, left it unclear whether and how the new NSF would support the social sciences.

As the first chapter's consideration of the NSF debate indicates and as subsequent chapters explain more fully, basic questions about the scientific identity, practical utility, and political import of the social sciences attracted extensive attention and provoked considerable controversy in the early postwar years. The second, third, and fourth chapters examine the stories of the military, the Ford Foundation, and the new NSF, respectively, to describe how each patron staked out its importance within the context of a transformed and largely new Cold War patronage system, to analyze the ways patrons and the scholars who worked most closely with them addressed long-standing questions and contemporary disputes about the social sciences and their funding, and to illuminate pointed challenges that arose as these patrons sought to advance scholarship grounded in scientistic and social engineering commitments.

By midcentury nobody doubted that the recently unified Department of Defense (DOD) was and would remain the dominant patron of American science for the foreseeable future. So for social scientists seeking support for their work in the Cold War years, the enormous defense science establishment naturally had great significance. Building on a sizable body of work about the military–social science partnership that includes many excellent accounts of specific disciplines and interdisciplinary fields of inquiry, chapter 2 focuses on the development of military funding policies and programs and examines the struggles of social scientists to establish their presence in the natural science–oriented defense science establishment. These scholars encountered persistent conservative suspicions and often found it hard to gain support from their superiors in the defense science establishment, including skeptical physical scientists. Under these conditions, social scientists had little choice but to argue strongly for scientistic forms of inquiry and their social engineering applications. Such ideas then became pervasive in military social science agencies and programs, thereby providing valuable support to many influential fields of research in ways consistent with those social engineering commitments. Moreover, these developments stimulated the growth of the military–social science partnership, which became increasingly important to American military operations and Cold War strategy by the time of the Kennedy administration.

At the middle of the twentieth century, the modern Ford Foundation was on the verge of becoming the largest and most influential private source of funding for the social sciences. At that time, the military's involvement with the social sciences remained rather underdeveloped, and the Rockefeller and Carnegie philanthropies themselves were struggling to devise a coherent plan for the social sciences. In this context, the Ford Foundation quickly assumed a major role in promoting what Ford personnel and documents commonly referred to as the behavioral sciences, which Ford's new Behavioral Sciences Program (BSP) became responsible for supporting as a scholarly enterprise.

As previous historical accounts have noted, the BSP and some other Ford programs provided vital support in the development of the behavioral sciences as a broad scholarly movement in midcentury America. This movement championed scientistic and social engineering viewpoints. It influenced many specific fields such as development studies and behavioralism in political science. And it provided valuable resources for action-oriented programs that shaped major policy initiatives during the Kennedy and Johnson administrations. Chapter 3 provides the first detailed reconstruction of the BSP's conflict-ridden development and the associated struggles of Bernard Berelson, the program's first and, as it turned out, only leader. As Berelson sought to fulfill the program's mandate to advance the scholarly resources of the behavioral sciences, he confronted mounting political pressures, in the form of McCarthyite attacks, conservative critiques of scientism, and growing internal discontent because of some Ford trustees' doubts about BSP's scholarly orientation and the public relations risk it posed. BSP's rocky history thus reveals as much about certain vulnerabilities in the new patronage system as it does about the pervasive appeal of the unity-of-science viewpoint within this system.

Chapter 4 returns to the case of the NSF, which acquired increasing importance as a major patron of American science and especially science carried out by university scholars. Founded with a special mandate to promote basic science, the NSF became involved with the social sciences slowly and cautiously. Based on carefully crafted policy principles first put forth by the sociologist Harry Alpert, who served as the agency's first social science policy architect, the NSF funded work only at the so-called "hard-core" end of the social research continuum, with an emphasis on basic research (rather than applied studies) steeped in quantitative analysis and committed to a value-neutral, investigative stance. Gaining a foothold in this natural science–oriented agency provided a major boost to the social sciences' scientistic wing, as a few historical accounts written mainly by social scientists, especially sociologists, have recognized. This chapter analyzes the NSF's development as a major patron for the social sciences, including Alpert's crucial role in developing a program for supporting hard-core research, the gradual expansion of this program within certain well-defined boundaries, and the difficulties that Alpert and other advocates of the social sciences encountered as they tried to convince the agency's natural science–oriented leaders to develop a more robust social science program. Contrary to the standard interpretation of Alpert, this chapter shows that he expressed some serious reservations about the narrowness of NSF's scientistic approach and the scientific pecking order that supported it. Consideration of some other critiques of NSF's carefully circumscribed program reveals additional challenges to NSF's scientistic strategy and the social sciences' marginal position within the agency.[10]

In all these chapters, efforts to advance the social sciences based on a unity-of-science viewpoint depended heavily on what science studies scholars refer to as "boundary work." Philosophers of science have proposed demarcation criteria for distinguishing science from nonscience typically by focusing on proper principles of scientific reasoning and scientific methodology. Historians of science and sociologists of science have examined the challenge of demarcation, but more so in terms of social practices and rhetorical strategies. While the practices and strategies of all these scholars may draw on philosophical discussions, they take shape through the efforts of a wide range of participants acting in various contexts who influence the direction, status, and use of social science research. These participants include natural scientists and social scientists, university administrators, leaders and program officers at public agencies and private foundations, politicians, and sometimes other parties who seek to establish the relationship of the social sciences to other spheres of human activity.[11] In the present study boundary work plays a pervasive role in the debates about the social sciences, in the development of patron policies and programs that promoted scientistic and social engineering commitments, and in the efforts that challenged those commitments.

Of course the period marked on the one end by Bush's 1945 science policy report and on the other by John F. Kennedy's presidency was not the first in which social scientists acquired significant support from patrons in the public and private sectors. Nor was this era the first in which extensive boundary work informed the directions of funding, which, in turn, influenced the production of social and psychological science knowledge, shaped the relationships between the social and the natural sciences and between the social sciences and the humanities, and contributed to the practical uses of social science knowledge. How, then, should one understand the significance of the new patronage system in the longer history of the social sciences in the United States?

Scientific Identity, Social Utility, and National Needs in a New Age

Are the social sciences really sciences? Is it possible and desirable to study human beings and societies in ways that are basically similar, perhaps even identical, to those found in the natural sciences? Do the social sciences acquire their relevance to practical affairs in the same ways that the natural sciences do? Scholarly works on the development of the modern social sciences since the seventeenth century indicate that no questions have been so fundamental and so problematic as these.[12] Some critics have doubted that the social sciences could ever rise above their status as "soft" sciences or, according to their harshest detractors, pseudosciences. On one occasion, Edward Teller, the physicist and "father" of the hydrogen bomb, declared that social science was no more scientific than Christian Science![13] Recently, it has also been said that efforts to make the study of economics more scientific and more useful through economic forecasting have helped—albeit unwittingly—to make astrology appear more respectable. Such damning quips are not unusual, and they rest on deeper suspicions about the very idea of creating a "science" of psychological, cultural, social, economic, or political phenomena. Naturally, such suspicions have encouraged proponents of these sciences to devote extensive time and energy trying to establish the scientific legitimacy of their work as well as its practical value, including its ability to advance the national welfare by addressing pressing national needs.

In the story of American social science, the interrelated issues of scientific identity, social utility, and national needs received much attention during the period of transition from an amateur- to a professional-based social science enterprise. The late nineteenth and early twentieth centuries saw the rise and demise of the amateur-led American Social Science Association (ASSA) alongside the development of the separate, and as it turned out more stable, professional social science associations for economics, sociology, political science, history, psychology, and anthropology. The processes involved in establishing national associations and university-based professional disciplines also involved extensive boundary work. As important historical works have shown, from the 1880s through the 1920s, scholarly leaders of the new professions worked hard to separate their work from other pursuits that threatened to embroil individual scholars, their research, and their scholarly institutions, including the universities, in destructive ideological and political disputes. Thus over time many leaders of the newer professional disciplines came to frown on the easy mixing of scientific social inquiry with the value-laden missions of social reformers, which had been common in the ASSA as well as in some community-oriented reform projects established during the Progressive Era, including Chicago's famous Hull House led by Jane Addams.[14]

During the 1920s the large private philanthropic foundations associated with the Rockefeller and Carnegie fortunes contributed to this effort to separate professional social science and scientific inquiry on the one hand from social advocacy and social reform on the other. These foundations did not abandon the project of linking social science to social progress though. Instead, they expected the new professions to contribute objective and apolitical knowledge needed to diagnose, understand, and manage the major social problems of the day, such as the continuing conflicts between labor and capital that contributed to economic instability and social strife. With this understanding, Rockefeller and Carnegie provided major funding for scholars at leading universities, including Chicago, Columbia, Harvard, and Johns Hopkins, at new social research institutes, such as the National Bureau of Economic Research and the Brookings Institute, and at the new SSRC.[15]

This picture sketched above, in which the social sciences became deeply committed to a broad-church scientism, to related investigative ideals including value neutrality, nonpartisanship, and objectivity, and to a complementary social engineering perspective has been prominent in scholarly accounts. In her landmark book *The Origins of American Social Science*, Dorothy Ross placed the developments summarized above in a rich intellectual narrative about the changing relationship between the social sciences and American liberalism. According to her analysis, scientism and its association with a managerial, technocratic approach for dealing with the nation's social problems became dominant by the end of the 1920s. Moreover, she claimed that "1929 takes us far enough into that history to recognize both the continuing characteristics of American social science and its distinctive twentieth-century features."[16] In more recent works, this viewpoint continues to hold great sway. For example, in his perceptive historiographic essay about the human sciences in Cold War America, the British intellectual historian Joel Isaac writes that the "growing secularization and professionalization of the university during the first half of the twentieth century was already inducing interests in technical, scientistic, and universalistic forms of inquiry." Certainly, historians of the Cold War years should recognize that these forms of inquiry had deep roots in earlier developments.[17]

Yet I believe that the picture sketched above requires important adjustments, for a few reasons. For one, serious challenges to the scientistic project arose during the interwar era and especially during the 1930s. As some convincing studies have shown, critics of that project found its investigative ideals wanting. They raised doubts about the stance of the supposedly objective scholar who could offer technical guidance to help manage social ills but who would, at the very same moment, remain disinterested in the sense of not making value judgments about what should be done. During a time when ominous developments at home and abroad threatened the future of capitalism, democracy, and freedom, the investigative ideals of value neutrality and nonpartisanship received sharp criticism from prominent scholars and public intellectuals associated with a variety of different ideological and political positions. During the 1930s, conflict over basic questions concerning the nature and uses of the social sciences threatened to split the major professional disciplines into hostile factions as well.[18] Furthermore, discussions at the Rockefeller and Carnegie foundations during the 1930s and continuing into the 1940s revealed serious disagreements on these same issues, a point considered more closely in chapter 3 because it provides crucial context for understanding the Ford Foundation's involvement with the behavioral sciences.

On top of this, as the social sciences became more directly involved with national politics starting in the 1930s, the boundary between the scholarly and the political became increasingly difficult to discern. Economists and other social scientists contributed to a wide range of New Deal policies and programs, to government planning and propaganda efforts during World War Two, and to postwar initiatives that aimed to extend New Deal–style government programs and planning. Under these conditions, was it reasonable to believe that social scientists and their research would remain objective and value neutral? Not according to many unhappy and often outraged conservative figures in the intellectual, business, and political communities who claimed the American way of life was now under threat from left-leaning professors.[19]

In short, though the scientistic outlook surely became widespread during the interwar era, that outlook also came under serious attack from various viewpoints within scholarly, philanthropic, and political circles. Following World War Two, the challenge of determining the character and future development of the social sciences resumed in earnest. At this critical juncture, the emergence of new sources of patronage acquired special prominence in the efforts to work out the meaning of the social sciences in the nuclear age.

The new patrons examined in this book and their social science collaborators all supported an approach that recognized the social sciences as junior partners in a scientific enterprise led by natural scientists. Some critics continued to deride this approach as unreasonably limited in its outlook, and some characterized it as foolhardy scientism. But its proponents presented it as the only truly scientific approach. They also argued that it provided a solid basis for advancing the social sciences' practical utility. Much as natural-science leaders claimed their work provided the foundation for medical, technological, commercial, and military applications of enormous value, social scientists and their patrons asserted that applications of their work would, in a parallel manner, contribute to American progress both at home and abroad, a point that brings to mind the political scientist Kenneth Prewitt's astute observation that the central project of American social science has always been America itself.[20]

During the Cold War years debates about the social sciences and their funding often moved back and forth between consideration of "academic" matters, for example, questions about the nature of scientific methodology and theory, and consideration of national needs and the type of knowledge that could best advance the healthy development of the United States as a dynamic modern society and leader of the free world. Social scientists and their patrons thus became involved in what we could call the co-construction of social science and society.[21] As the following chapters reveal, when patrons and social scientists tried to define the scientific character of this work, questions about what type of knowledge the nation needed in the anxious nuclear age were never far from view.

Thus by focusing on a small group of influential new patrons in the early Cold War decades, the present study deepens our understanding of the intertwined stories of scientific identity, social utility, and national needs in the evolution of the social and psychological sciences. Certainly, the complicated growth of professional social science in earlier decades provides crucial background. Yet, in the present study, understanding the evolution of patron support for scientistic and social engineering commitments requires careful attention to how patrons, social scientists, and other interested parties responded to prominent developments in American science, politics, and higher education during the Cold War years themselves. Those developments include the dramatic expansion of the federal science system and the defense science establishment; the powerful presence of the natural sciences and especially the physical sciences in federal and defense science agencies; bitter partisan debate about the legacy of the New Deal; the growth of anticommunist politics; the early postwar marginalization of left-liberal positions in the political, scientific, and academic communities; the rise and decline of McCarthyism; and the resurgence of liberal Democrats along with a more vigorous liberal reform agenda by the early 1960s. By situating new funding policies and programs in these contexts, this study illuminates key features of the politics–patronage–social science nexus and its evolution in Cold War America.

Contemporary Challenges and Later Critiques

The development of new patrons in a loosely coordinated funding system helped set the stage for fierce controversy about the character of the politics–patronage–social science nexus starting in the mid-1960s. During the Kennedy and Johnson administrations, some social scientists and their patrons, especially the military and the Ford Foundation, acquired considerable influence in the development of major new domestic, military, and foreign policy ventures. Yet by the mid-1960s and more obviously by the late 1960s, mounting criticism in the scholarly and political arenas raised troubling questions about the recently acquired national influence of the social sciences and their patrons. From a variety of viewpoints, a broad array of detractors attacked the scientistic and social engineering commitments as a sham. After working so hard to promote those commitments as the keys to scientific maturity, social utility, and public policy relevance, scholars and their patrons now found themselves under fire, considered by an array of critics to be part of the nation's explosive problems, not their solutions.

Elsewhere, I have argued that, starting in 1965, the controversy over Project Camelot, a U.S. Army–sponsored research project that promised to facilitate the development of effective counterinsurgency measures, provided a critical catalyst for debates over military funding of the social sciences. That debate, in turn, stimulated increased scrutiny of the politics–patronage–social science nexus more widely.[22] Though that analysis of Camelot's significance still seems correct to me, the present study indicates that the explosive controversies of the mid- to late 1960s were informed by the five earlier challenges to the Cold War funding system, which I mentioned before and which now require additional explication.

In many episodes and contexts, natural scientists, politicians, and intellectuals associated with various conservative positions raised suspicions that linked social scientists and their supporters to a variety of left-leaning and supposedly dangerous causes, thus constituting one main challenge to the new patronage system. In addition, regarding the question of scientific identity, conservative criticisms pointed in contradictory directions. Some conservative critics, including Vannevar Bush and other influential figures in the natural-science community, indicated that the social sciences, or at least certain parts of them, could become nonideological, apolitical, and rigorous in their investigations much as the natural sciences seemed to be. However, other conservative figures from the scientific, political, and intellectual spheres claimed that the social sciences had gone down the wrong path by trying to ape the natural sciences. Moreover, these critics charged that scientism harmed the moral, social, and political fabric of the country. Both lines of criticism surfaced during the postwar NSF debate and then became prominent during the McCarthy Era, when they became particularly troublesome for the large private foundations including Ford and for the NSF as it began to fund the social sciences.

Meanwhile, a small number of liberal scholars raised a second challenge as they too found the value-neutral, nonpartisan, and technocratic orientation of the social sciences troubling. Whereas conservative critics of scientism typically suggested that the pretense of scholarly objectivity and value neutrality hid leftist biases in the social sciences and thus obscured their roles in promoting such harmful developments as secularism and big government, liberal critics typically claimed that by trying to imitate the natural sciences, social scientists compromised their ability to critique the unjust status quo, undermined their potential to advance programs for progressive social change, and left themselves ill prepared to evaluate the threats posed by powerful interests eager to use their work for undemocratic purposes. In the immediate postwar years this position had its strongest advocates among left-liberal scholars, including the sociologist Louis Wirth, while in the late 1950s and early 1960s this position received support from scholars associated with the nascent New Left, including the sociologist Irving Louis Horowitz.

A third challenge concerned the more general threat that patrons would direct the course of scholarly research in unseemly ways. While conservative and liberal criticisms of scientism indicated the seriousness of this threat, one did not have to adopt either of those positions to realize that the power of the purse might have detrimental influences. Indeed, leading social scientists and other parties who advocated greater funding from Cold War patrons regularly acknowledged this potential danger, as evidenced by the attention they gave to theoretical discussions, institutional arrangements, and practical measures that could promote healthy relationships between scholars and patrons. According to a common historical interpretation, during the first decade and a half or so following World War Two social scientists made little effort to evaluate threats that patronage posed to the quality and independence of scholarship, especially from the military and intelligence agencies.[23] But the present study suggests that it is more accurate to say that social scientists generally expressed confidence that those threats could and would be managed effectively. Moreover, the challenge of preventing those threats from causing real damage remained ever present and only grew larger over time, as the importance of patron-funded work became more visible both inside and outside the ivory tower.

The second-class status of the social sciences in the Cold War patronage system presented a fourth challenge. Efforts by new patrons to stimulate the development of the social sciences took place within political, scientific, and institutional contexts where the natural sciences often represented the gold standard. Though natural scientists did not have such a powerful presence at the Ford Foundation, the planning and development of Ford's BSP reflected common criticisms of the social sciences from the nation's natural-science elite. Conditions inside the federal science establishment, including defense science agencies and the NSF, proved more problematic in the long run, as the higher status and greater influence of natural scientists allowed them to exercise strong oversight of social science funding and programs. As a result, not only did social science representatives and their allies find themselves constantly on the defensive, but some of them also reasoned that the reigning scientific hierarchy posed an obstacle to the healthy development of these sciences and their social contributions.

The fifth challenge arose from the blurring of boundaries between social scientific inquiry and value-laden analysis. Though the strategy of promoting the social sciences along scientistic lines supposed that the production of knowledge would remain objective and value neutral, the distinction between scientific inquiry on the one hand and political objectives, social reform, and democratic ideology on the other became rather indistinct in practice, at least in certain fields of patron-supported research.[24] The most obvious cases considered in this book involved the Ford Foundation and military agencies. They and their grantees often sought to address issues of great political and ideological importance, including questions about how to strengthen American democracy at home and how to undermine communist influence abroad. To be sure, research that addressed such questions often involved data collection, empirical analysis, and theoretical discussion without necessarily providing explicit prescriptions or policy recommendations. However, the frameworks that guided the conduct of such research as well as the anticipated uses of the resulting knowledge often rested on value judgments, social goals, and political commitments that even at the time were not at all difficult to spot.

As the discussion above suggests, the importance of each of the five challenges varied from patron to patron, and thus the challenges will receive different degrees of consideration in the following chapters. But when considered together, they reveal that the social science patronage system from the mid-1940s to early 1960s encountered significant problems and, to a greater extent than historical scholarship has typically recognized, rested on shaky foundations. This claim does not imply that the explosive controversies during the second half of the 1960s were inevitable. But it does suggest the need to situate those controversies against a complex historical background, recognizing that new patrons strongly promoted scientistic and social engineering commitments during the earlier Cold War years and that those efforts encountered important criticisms, obstacles, and setbacks. Subsequently, those challenges informed the development of vigorous political and scholarly controversy about the politics–patronage–social science nexus starting in the mid-1960s as well as corresponding efforts to transform that nexus in different ways. Though the present study cannot consider those later developments in any detail, in the conclusion I outline notable aspects of the mounting controversy during the 1960s, identify important efforts to reform patron engagements with the social sciences in subsequent years, and consider the meaning of those developments for understanding ongoing debates about scientific identity, social utility, and national needs.

A Note on Terminology

"Scientism" is a sensitive term. Historical figures who supported the general position described in this book as scientistic would not have put it this way, for scientism typically was a term of disdain—and it remains so today. The term's negative connotation has special relevance to the present study because one of the twentieth century's harshest critics of scientism, Friedrich Hayek, became well known during the early Cold War years. A leader of the Austrian school of economics, an aggressive advocate of free-market economics, and a future Nobel laureate in economics, Hayek put forth in his writings a scathing history of efforts to use the natural sciences as a model for advancing the social sciences. Tracing the roots of this impulse to the nineteenth-century work of French protosociologists Auguste Comte and Saint-Simon, Hayek emphasized that their enthusiasm for national planning formed part of a wider project to reconstruct society along scientifically rational and efficient lines. In Hayek's view, that project had disastrous consequences, as seen most clearly when latter-day enthusiasts of socialism, most notably in the Soviet Union, put its core ideas into action and thereby paved the way for horrendous assaults on personal freedom, runaway expansion of a repressive state apparatus, and frightening economic inefficiencies. Apart from this lineage connecting nineteenth-century scientistic schemes with twentieth-century socialist nightmares, Hayek claimed that there were fundamental differences between the natural and social sciences at many levels, regarding their proper objects of inquiry, methodologies, goals, and social uses. Similarly, other writers in the mid–twentieth century who employed the term scientism typically did so in order to criticize the extension of a natural-science outlook to the social sciences or other areas of inquiry, like philosophy.[25]

Nevertheless, one can use the term scientism for the purposes of description and analysis, rather than prescription or pejorative commentary. In a recent book about scientism in nineteenth-century European thought, Richard Olson deploys the term effectively for the purpose of identifying a widespread though also contested viewpoint about the nature of social science inquiry and progress. In doing so, Olson examines a variety of efforts to use natural-science methods, concepts, and theories in the social sciences, but he does not do this in order to criticize those efforts. By pointing out that they all had a scientistic character, he does not imply they were therefore fundamentally flawed. Neither does he fail to recognize important differences among those efforts, even though they all shared something important in common.[26]

Similarly, in this study, I use the term scientism for descriptive and analytic purposes. However, I do not mean to hide my personal views from the reader. So I want to make it clear that I believe many types of worthwhile social science inquiry do differ in basic respects from natural-science inquiry, for example, in physics, chemistry, cosmology, geology, and botany, because only practitioners of the former need to understand, and thus need to develop specialized investigative tools for understanding, such things as ideas, emotions, religion, culture, social norms, social order, politics, and justice. Still, when I say in this study that some individual or patron or field of research advanced a scientistic outlook, I am stating what I take to be a fact supported by the evidence I have examined, not implying a criticism of that outlook. And I have tried my best to characterize fairly the positions of those who favored scientistic approaches.

The choice of terminology for discussing the social sciences can also be a sensitive issue. In the last few centuries, thinkers have proposed various terms referring to the project of studying human beings and social phenomena in a scientific manner. In addition to the common phrase social science, one finds such terms as the science of man, cultural science, behavioral science, and human science, as well as their plural forms. Various terms have been used to refer to different views of more specific areas of study as well. For example, the study of political phenomena has been called government, political studies, and political science. The proliferation of these terms is revealing, for it reflects the persistence of long-standing controversies about the nature of this broad area of study and its many regions. For instance, advocates of the term political studies typically did not believe it was desirable or even possible to pursue the study of political phenomena in a manner that really qualifies as scientific. From the other side, those who promoted the term behavioral sciences often expressed strong confidence in the project of studying social and psychological phenomena in a rigorous scientific manner that had much in common with the natural sciences.[27]

But, once again, my purposes here are not prescriptive. Thus I have chosen my terms in order to convey this study's descriptive and analytic claims as clearly as possible, rather than to advocate or criticize the use of one term or another. I often use the common and rather generic terms social science, social sciences, or social and psychological sciences in order to refer to this broad area of study. When referring to specific disciplines or specific fields of study, I usually rely on the terms most common at the time and whose meaning in a broad, everyday sense still remains much the same, such as psychology, sociology, and economics. In some places I use other well-known terms, including the behavioral sciences, because I find them especially appropriate to my historical discussion at those points. Sometimes I also invoke less commonly used terms, such as the sciences of choice and the science of control, but only when they seem needed for reasons I hope will be clear to the reader. In order to refer not only to the ideas, research, theories, and aims of the social sciences but also to the people, interests, and organizations involved, I often find the phrase social science enterprise helpful.

*****

In *The Proper Study of Mankind*, a 1948 book about recent advances in the social sciences that sold extremely well, the journalist Stuart Chase told the following anecdote. After the first atomic bombing of Hiroshima on August 6, 1945, it seemed clear that something of grand and disturbing importance had occurred.

It seemed equally clear that discerning the implications of that dramatic event for the future of humanity would be a complicated task. In their search for insights, journalists immediately turned to physicists and other natural scientists. But nobody thought to consult a social scientist, at least not until day three.[28] Though perhaps apocryphal, Chase's anecdote cleverly captured a widespread lack of confidence in the social sciences, sometimes accompanied by a marginal interest in them, on the part of many politicians, scientists, and opinion makers at the dawn of the atomic age. At the time, prominent social science scholars knew the truth of Chase's anecdote all too well. Not surprisingly, they also worked hard to convince those skeptics that they too had a great deal to contribute to the nation's well-being. As discussed in the next chapter, the postwar NSF debate provided an early focus for efforts to define the character, future development, and public funding of the social science enterprise in the context of major transformations in the nation's scientific and political landscapes.

1

Social Science on the Endless (and End-less?) Frontier

The Postwar NSF Debate

If social scientists aspire to the status and position and public estimation of other scientists, they must subject themselves to standards of the kind recognized by other scientists and by the public. That is, they must specify criteria that distinguish social scientists from that vast array of camp followers, reformers, propagandists, and social workers, which today dominate even most of the professional organizations of social scientists.

—Sociologist George Lundberg, 1947

No amount of aping of the methods of the natural sciences or attempt to bask in the reflected prestige of the natural sciences will be sufficient.

—Sociologist Louis Wirth, 1947

World War Two marked a major turning point in the relationship between the federal government and American science. During the war, annual federal support for scientific research soared, from $48 million to $500 million, from 18 percent to 83 percent of the nation's total science funding. Even before the war's end American leaders commonly believed that, henceforth, national defense and public welfare would depend on a greatly expanded national commitment to science. The success of the Manhattan Project and the role of atomic weapons in bringing about a sudden Japanese surrender helped solidify public support for a robust postwar scientific enterprise supported by federal funds.[1] Under these conditions, the nation's scientists joined politicians and prominent figures from business, labor, the military, higher education, and other sectors of American society to consider proposals for a major new science agency, a National Science Foundation (NSF). Following the presentation of Vannevar Bush's science policy report, *Science—The Endless Frontier* (*SEF*), to President Truman in July of 1945, legislators put forth competing bills. One proposal associated with a conservative science policy agenda drew heavily on the viewpoint presented in Bush's report, while Senator Harley Kilgore presented an alternative proposal closely tied to a liberal science policy agenda.

The ensuing national science policy debate, which would ultimately involve more than twenty legislative bills, continued for a few years and thus took place during a crucial transitional moment in world history, in American history, and in the history of American science. In international affairs, a terrifying Cold War replaced a brutal hot war. By midcentury, the United States, leader of the democratic, capitalist Western bloc, had locked horns in a seemingly unending struggle with the Soviet Union, head of the totalitarian, communist Eastern bloc. As the superpower conflict deepened, it also spread to all regions of the globe. In domestic politics, postwar partisan conflict raged between liberals interested in extending the legacy of the New Deal and conservatives determined to put an end to so many years of "treason." In addition, a widespread desire for normalcy coexisted uneasily with a pervasive fear that the so-called American way of life faced grave dangers, not just from communists abroad, but also from a vast left-wing conspiracy inside the country's borders. Against this complex background of international conflict and domestic strife, the NSF debate became a pivotal episode not only in the origins of the new science agency but also in the development of postwar national science policy more widely.[2]

This debate also sparked controversy over the question of whether or not to include the social sciences. Though neither *SEF* nor the initial legislative bill based on Bush's report included the social sciences, in the fall of 1945 Senator Kilgore and the Truman White House expressed interest in including them. Kilgore also invited the Social Science Research Council (SSRC) to present the position of the social sciences at legislative hearings. Participants in the NSF debate thus had an important opportunity to express their views about the scientific status, social purpose, and national contributions of the social sciences at a critical historical juncture.[3] My analysis of the controversy over the social sciences highlights three developments that had deep significance for the intertwined debates about their scientific identify, social uses, and political relevance.

First, this controversy revealed that the political and science policy landscapes of the early postwar years posed significant difficulties for social scientists and their supporters. Not only did this debate help to establish the position of the social scientists as second-class citizens in the emerging postwar federal science system; it also galvanized a powerful and mainly conservative coalition of natural scientists and politicians who supported legislation inspired by *SEF* and who expressed strong reservations about the social sciences. Meanwhile, another group of mainly liberal scientists and politicians supported inclusion of the social sciences through Senator Kilgore's legislation, which naturally reinforced conservative suspicions about the social sciences' leftist tilt.

Second, under these difficult conditions, a group of prominent SSRC scholars led by the economist Wesley Mitchell developed a case for inclusion based on a scientistic viewpoint. During the interwar era, SSRC scholars had been deeply involved in the debates over advocacy and objectivity. And in the early postwar years some social scientists on the liberal left, including the sociologists Robert Lynd and Louis Wirth, continued to criticize the effort to take the natural sciences as an investigative model. But the SSRC group led by Mitchell argued in their congressional testimony that social scientists should position themselves as part of a unified scientific enterprise. Accordingly, they all accepted a sharp distinction between objective social research on the one hand and social reform, political ideology, and value-laden inquiries on the other hand.

Third, the course of the NSF debate left questions about the social sciences' scientific status and political support just as problematic as they had been at the beginning of the debate. Though the SSRC group worked hard to convince their critics that their work had a great deal in common with the natural sciences, skeptics from conservative political and scientific circles remained fiercely suspicious, claiming the social sciences exhibited a troubling immaturity in comparison to the natural sciences or, more forcefully, rejecting the very notion that one could ever study human nature and society in a rigorously scientific manner. Moreover, as we will see, the strongest political and scientific detractors of the social sciences managed rather quickly to place them on the sidelines of the ongoing NSF debate.

An Ominous Beginning

Our story begins with the social sciences offstage. This point is crucial, because in this episode and in the years to come, American social scientists were often scrambling to respond to transformations in science, politics, and science policy controlled mainly by other, more powerful actors in the political and scientific communities. By the time social scientists became directly involved in the postwar NSF debate, this debate had already acquired a decidedly partisan character. Though one cannot draw a simple boundary between liberal and conservative positions on federal science policy that neatly fits every participant on every occasion, this debate pitted a conservative agenda most closely associated with Vannevar Bush against a liberal agenda promoted by the West Virginia Democratic senator Harley Kilgore. These and other influential non–social scientists often viewed the social sciences in relation to their partisan agendas, with Bush and his allies in the scientific and political communities trying to keep the social sciences out of the proposed agency, while Senator Kilgore and the Democratic White House led by President Truman sought to include them.

With strong connections to organized labor and the New Deal wing of the Democratic Party, Senator Kilgore chaired a Senate subcommittee that he used to advance a liberal postwar science policy agenda. In Kilgore's view, the national welfare would be best served if a new, comprehensive science agency developed a coherent set of national research priorities; if the public had free access to the results of publicly funded research; if federal patronage supported research in universities from a wide range of geographical locations; and if the proposed agency was politically responsive to a variety of interest groups, by having representatives from small business, labor, and the public on the agency's governing board and by having the president appoint its director.[4]

Vannevar Bush, leader of the conservative countereffort, was an electrical engineer by training, a former dean of engineering and vice president at MIT, a former president of the Carnegie Institute of Washington, and the director of the central wartime science agency, the Office of Scientific Research and Development (OSRD). Bush abhorred the centralizing tendencies of the New Deal and the expansion of federal power into areas of American life previously handled by the private sector, philanthropy, and local government. In *SEF* Bush, like Kilgore, called for the creation of an "over-all" science agency. But therein ended the similarities.[5] In contrast to Kilgore, Bush proposed that to preserve "freedom of inquiry," the government had to allow institutions that carried out scientific research to maintain "internal control of policy, personnel, and the method and scope of research"; that patents deriving from agency-sponsored research should not become governmental property; that the agency should not be required to follow any rule concerning the geographical distribution of public funds; that science should be kept separate from politics by having the agency's governing board composed of private citizens including, presumably, many scientists; and that this board of scientifically minded private citizens, not the president, should choose the agency's director. Above all, Bush emphasized the need to keep science free from political control.[6]

Soon after Bush presented *SEF* to Truman, Congress unveiled competing proposals for a new science agency. One proposal came from Washington State senator Warren Magnuson, a New Deal Democrat and friend of President Truman. In these respects Magnuson had much in common with Kilgore. However, Magnuson, who had a friendly relationship with Bush, introduced a Senate bill that followed Bush's conservative plans to a tee. Wilbur Mills, a Democrat from Arkansas, introduced a companion House bill. Furious, Kilgore reasserted himself by introducing an alternative Senate bill based on his preferred policy principles.[7]

Up until this time the social sciences were noticeable by their absence. Kilgore's bill mentioned only the natural sciences and "related economic and industrial studies," probably out of an overriding concern with the natural sciences rather than an impulse to exclude the social sciences specifically. In *SEF* and associated legislative proposals the social sciences received no consideration, though in *SEF* Bush noted that they should not be overlooked: "it would be folly" to ignore "the social sciences, humanities, and other studies so essential to national well-being." In a letter of transmittal accompanying *SEF*, Bush explained to Truman he had excluded the social sciences from the proposed agency because former president Roosevelt's 1944 request for a national science policy report "had in mind the natural sciences, including biology and medicine."[8]

However, Bush had helped to design Roosevelt's request so that it would exclude the social sciences, though he did not tell Truman this. According to the sociologist William Ogburn, when Roosevelt had asked Bush to prepare a letter inviting Bush to conduct his science policy study, Roosevelt had specifically said that the social sciences should be included. Subsequently, Bush omitted the social sciences in his letter, but Roosevelt did not comment on this omission when he signed it.[9] Furthermore, in private correspondence Bush noted he had "made sure" Roosevelt's letter "confined its attention to the natural sciences."[10] Bush had at least three good reasons for believing his scientific and political allies would not want the social sciences included.

First, an important segment of the nation's natural-science elite had never recognized the social sciences as equal partners. Consider the case of the National Academy of Sciences (NAS), where Bush had been a member since 1934. A private, nonprofit society established by an act of Congress during the Civil War, the NAS provided advice on scientific matters to the federal government, carried out studies requested by federal agencies, and brought together leaders from government, industry, and academia to assess the state of scientific inquiry in fields relevant to national interests. But from the social sciences only anthropology and psychology had regular representation within NAS. Furthermore, academy leaders had not found it "easy to set the boundaries for anthropology and psychology within the framework of the parent body, representing as it does by tradition the natural sciences," explained an internal 1954 NAS report. Thus the academy's work in anthropology and psychology remained limited to their physical or biological aspects through the early post–World War Two years.[11]

National science policy events during the Great Depression had contributed another layer of friction between social and natural scientists. During that crisis, eminent social scientists associated with the SSRC including the

University of Columbia economist Wesley Mitchell and the University of Chicago political scientist Charles Merriam sat on the governing board of a controversial New Deal planning agency. First established in 1933 within the Public Works Administration, this agency (after a number of organizational and name changes) was called the National Resources Planning Board (NRPB), and in 1939 it became part of the president's Executive Office. In the mid-1930s the NRPB reviewed plans for a major federal science program proposed by a separate natural science–oriented agency called the Science Advisory Board (SAB). Created in 1933, merely two weeks after NRPB's establishment, the SAB had close ties to the NAS and included many of Bush's close scientific colleagues, such as the physicist Karl Compton, who became SAB chairman, while Bush himself chaired SAB's committee on patents. Not surprisingly, the two New Deal agencies clashed: the NRPB rejected the plan put forth by the SAB; SAB's death soon followed; adding insult to injury, the NRPB, with President Roosevelt's approval, then created its own advisory science committee, which produced *Research—A National Resource* (1938–1941), a landmark three-volume report on the federal government's scientific programs.[12]

Second, World War Two had shifted the balance of political influence dramatically in favor of natural scientists, especially physical scientists with NAS connections who dominated top-level positions at Bush's OSRD and starred in the major wartime projects that produced radar, the proximity fuse, computers, and atom bombs. Though social scientists made various wartime contributions, they were overshadowed by the wizardry of the "hard sciences." Moreover, social scientists had no central wartime agency comparable to the OSRD.[13] Meanwhile, with the exception of OSRD's panel for applied psychology and the presence of some economists in other OSRD panels or projects, this key node of wartime science largely excluded the social sciences.[14]

Third, because of social scientists' extensive involvement in New Deal agencies and policy initiatives involving social welfare, agricultural policy, financial regulation, labor relations, social security, and economic planning, the social sciences faced repeated and often brutal criticism from conservatives in the scientific, business, and political communities. The NRPB provides an illustrative example. Over a ten-year period, this social science–led planning agency developed an extensive philosophy of federal social insurance with cradle-to-grave welfare programs. What Charles Merriam and others called "A New Bill of Rights" included the right to decent work and fair pay; adequate food, clothing, shelter, medical care, education, and security; a system of free enterprise; equality before the law; and rest, recreation, and adventure. As the nation's energies turned to fighting World War Two, a coalition of outraged congressional conservatives voted against any further public appropriations for the NRPB. Consequently, the social science–led board expired in 1943.[15] By the mid-1940s, then, "intellectuals concerned with 'human affairs' in general" found themselves "in a less secure status than the physical and biological scientists who affect public policy," as the sociologist Robert Merton put it.[16]

Though prudence encouraged Vannevar Bush to try to keep the social sciences out of the postwar science debate, his own suspicions about this field gave him additional reasons. Some historical accounts suggest that in leaving out the social sciences, Bush acted on personal beliefs. Daniel Kevles has written that Bush saw social science as "so much political propaganda masquerading as science." Indeed, Bush, who found the social sciences' New Deal connections troubling, could be quite biting, claiming on one occasion that "a tremendous number of charlatans work under the social science tent."[17] But Bush did not express an across-the-board antipathy. At one point many years later, he even declared he was "very avid to see the social sciences" acquire "a sounder basis and become more useful."[18] Yet in the postwar NSF debate he never pressed for their direct inclusion, preferring to defer consideration of them until a later date.[19]

Despite Bush's efforts, however, the social sciences soon came into the picture "with a bang." That characterization came from Irvin Stewart, Bush's OSRD executive secretary, who referred to the president's message to Congress in September of 1945 in which Truman outlined a reconversion program from wartime to peacetime activities, urged speedy passage of Kilgore's science agency legislation, and called for the social sciences' inclusion. On this last point, Truman followed his Bureau of the Budget director, Harold Smith, who had warned him that if Bush's camp had its way, the nation might end up "with an unbalanced research program . . . with the social sciences being overshadowed by the impact of the atomic bomb."[20]

More intriguing, Truman may have also been considering the intentions of his White House predecessor, who had often turned to social scientists for advice. In his 1944 State of the Union address, President Roosevelt had advocated NRPB's new Bill of Rights. In addition, Roosevelt's last written words from his undelivered Jefferson Day address, scheduled for April 13, 1945, the day before Roosevelt died in office, include this striking passage: "Today we are faced with the preeminent fact that, if civilization is to survive, we must cultivate the science of human relationships—the ability of all peoples, of all kinds, to live together and work together, in the same world, at peace."[21] So Truman could have reasonably concluded that Roosevelt would have favored including the science of human relationships—as indeed Roosevelt did, if Ogburn's account is accurate—and thus decided to do the same.

In any case, Truman's reconversion address cemented the connection between the social sciences and liberal postwar science policy plans, a connection bound to provoke a harsh response from conservatives. In his 1992 book about NSF's engagement with the social sciences, the sociologist Otto Larsen has proposed that the problems faced by social scientists during the NSF debate lay in their isolation from government.[22] However, it is more accurate to say, as some other authors have noted, that the social sciences' pronounced involvement with the New Deal and its extensions during World War Two and the early postwar years fueled partisan controversy and anti–social science sentiment.[23] As we will see below, partisan conflict would pose a serious obstacle to social scientists' quest for a position in the proposed science agency.

Meanwhile, social scientists were discussing these fast-moving developments among themselves. Deliberations on postwar federal funding conducted by the SSRC became crucial, for the council would serve as the official voice of American social science in this episode. With respect to the proposed comprehensive science agency, the council first needed to determine whether or not to seek inclusion. As early council discussions would indicate, the prospect of greatly increased federal funding comprised a threat, not only an opportunity.

Social Scientists Take Stock: Public Funding as Threat and Opportunity

A number of considerations prompted social scientists to look to the SSRC for leadership in the NSF debate. Ever since its founding in 1923, the council, with its headquarters in New York City, had provided a central meeting place for prominent American social scientists from the various disciplines. The council thus became an especially valuable resource whenever social scientists confronted matters of national importance that cut across the intellectual and organizational boundaries separating sociologists from political scientists, or economists from psychologists. Additionally, the council promoted public appreciation of social science and advocated greater use of social science expertise in government. During the Depression Era, SSRC committees in such areas as social security and public administration worked closely with New Deal initiatives. In the early 1940s, the council set up an office in the nation's capitol in order to facilitate social scientists' wartime efforts. At that time, the council's basic composition and structure included a chief executive (or president); members representing the major national professional societies of psychologists, sociologists, political scientists, economists, anthropologists, statisticians, and historians; additional members at large; a central Problems and Policy Committee; and a group of standing committees to deal with specific topics of interest.[24]

In the most extensive history of SSRC's development through World War Two, the sociologist Donald Fisher contends that during the course of the NSF debate, "the SSRC never convinced itself that government support was essential for the growth of science." Fisher also states that a widespread fear that governmental support would corrupt scholarly research and undermine scientific objectivity became a "major cause" within the council, thus contributing to the elimination of any direct mention of the social sciences in NSF's 1950 enabling legislation.[25] As discussed below, internal deliberations among the council's social scientists did give expression to significant concerns about the power of the federal patron to direct the course of social research in worrisome ways. Such concerns received ample elaboration through the work of the council's Committee on the Federal Government and Research, chaired by Wesley Mitchell, a world-renowned economist and former member of the now-defunct NRPB. But we will see that as the council's deliberations progressed, they underscored the growing conviction among leading social scientists that postwar federal funding would be vital to their future growth, prosperity, and influence.

In the spring of 1944, SSRC discussions indicated that "no other single activity" had significance "immediately equal" to that of the relationship between the federal government and the social sciences. The council thus set up the new committee chaired by Mitchell.[26] The committee also included three scholars whose views we will consider below: the economist Edwin Nourse, the sociologist William Ogburn, and the psychologist Robert Yerkes. As they considered dramatic wartime changes in the relationship between American science and the federal government, the committee members became worried about the threat to scholarly integrity posed by increased federal funding, a worry with deep roots in the council's own history.

Earlier in the century, when private philanthropy took its first steps toward the support of scholarly social research, critics of big business warned that such funding came with the imprint of corporate capitalism, the source of philanthropic largesse, and would thus compromise the intellectual integrity and social respectability of scholarly inquiry. In response, the Rockefeller and Carnegie philanthropies, together with leading social scientists at the SSRC and elsewhere, defended nonpartisanship and value neutrality as scientific ideals. But, as noted in a 1937 SSRC report written by the Chicago sociologist Louis Wirth, during the Great Depression foundation support became "precarious and temporary." Furthermore, foundations increasingly focused on research that addressed practical problems. Trends in private funding thereby threatened to diminish the council's "independent judgment," dampen its "initiative," and accentuate "a policy of drift and opportunism," wrote Wirth with concern.[27] In 1944 C. Wright Mills, a young sociologist at the time, warned more generally that since private foundations viewed scholars who advanced "unpopular theses" with suspicion, social researchers supported by foundation money often censored themselves by choosing "safe problems in the name of pure science."[28]

In the mid-1940s, SSRC's Committee on the Federal Government and Research had to consider the growing threat posed by public funding, as unprecedented numbers of social scientists conducted research for federal agencies and as the prospect of greatly expanded postwar federal support for university research became clearer. The "pre-war balance between public and private research" would probably not be restored when peace returned, observed Mitchell's committee.[29] Above all, committee members feared that recent expansion of social research within federal agencies would result in significant damage to the social science enterprise. In a 1944 paper, the committee warned that in certain fields of social inquiry, "intellectual hegemony" might come to "rest with the respective research staffs of the Federal bureaus." This situation could, in addition, foster a more global "devaluation" of the "standards of work and of the credence given to the concept of research itself." In the long term, the consequences could be alarming, as "an entire generation of research workers" might suffer from a "permanent impairment of skill, objectivity, and interest."[30]

One year later such fears deepened within the SSRC. Anticipating the public presentation of Bush's science policy report (but not anticipating that he would choose to exclude the social sciences), Mitchell reported that his committee firmly opposed the "general subsidization of research institutions" by the federal government. In accord with wider discussion among the council's leadership, the committee argued that "a subsidy by the national government to universities and other institutions for research in general and not merely with respect to its own immediate needs, will not serve research or the national interest." In light of "the great importance of maintaining the social scientist's complete independence with respect to many delicate problems of government," federal subsidies would present "obvious special difficulties."[31]

Yet the council's scholars did not want to isolate themselves in a lonely, impoverished corner. Minutes from the powerful SSRC Problems and Policy Committee claimed that social scientists had a "universal desire for more funds." Thus "a negative statement" regarding federal support might "not be popular with the Council's constituents." In addition, the Problems and Policy Committee warned that a negative statement on federal funding could "be construed as prejudicial to the interests" of natural scientists, which "would be unfortunate."[32]

After Congress unveiled competing legislative proposals in July of 1945, the concern about political subordination receded further at the council, while the fear of being left out of the proposed centerpiece of postwar federal science grew stronger. Considering whether the SSRC should issue a public science policy statement, the Problems and Policy Committee reasoned that constructing a satisfactory statement would be "hardly possible," for "discussions of committee and Council members have shown shifting views." "Even the original consensus against government subsidization of private institutions for research in general seems no longer to prevail." Some social scientists, believing that previous sentiment had been "too hesitant and fearful," now wanted an "aggressive policy" favoring federal support. The Problems and Policy Committee added that it would be possible to devise ways for obtaining support while avoiding the "disadvantages from centralized control."[33]

This crucial shift in opinion hardened following Truman's September reconversion address, in which the president blessed the social sciences with his support. Among SSRC scholars a new consensus emerged on the point that increased federal funding, characterized as "a question of the gravest importance" at a board of directors meeting, would stimulate the social sciences in positive ways. A number of practical concerns contributed to this consensus. The University of Chicago demographer Philip Hauser warned that a new agency dedicated only to the physical sciences would drain valuable personnel away from the social sciences.[34] Mitchell's committee also recognized that the federal government represented the only main "source now in sight" capable of supplying the desirable level of "augmented funds" needed "if the development of research is to proceed at the maximum feasible rate."[35]

Brief consideration of other funding sources would have suggested their obvious limitations. Hard hit by the Great Depression, private foundations could no longer shoulder so much of the burden for supporting American social science. From industry, social scientists could expect funding for work related to commercial interests but little more. They could also expect the military to support studies related to its practical goals. But beyond that, the size and scope of postwar military social science programs remained uncertain circa 1945.[36]

Funding could have also come from a separate national agency for the social sciences. According to Harvard sociologist Talcott Parsons, following *SEF*'s publication "a considerable body of opinion," including a number of natural and social scientists, favored the creation of such an agency. However, social scientists soon concluded that Congress would not support such an agency, because their work was "politically controversial."[37]

It seems fair to conclude that practical considerations won over the council's social scientists. Their quest for inclusion in the proposed NSF became a pressing issue, overshadowing long-standing concerns about the impact of public funds on scholarly standards of work. Such concerns by no means disappeared though, as Mitchell's committee continued to emphasize the importance of "freedom of inquiry and the independence of the spirit of research." His committee still warned that "restraints imposed by policy and interest tend to debase research into the preparation of apologia and rationalizations for the courses of those under whose dominance research is undertaken. Social science, under these circumstances, ceases to be scientific in spirit save perhaps on the purely technical and methodological level. . . . a sorry form of social science." But Mitchell's committee and SSRC's social science representatives more widely had concluded that safeguards, like the development of diverse sources of support, could prevent an unacceptable degree of federal control. Equally significant, the SSRC almost seemed "forced to approve of social science participation or else lose any influence it might have," as the historian and eminent archivist Waldo Gifford Leland observed.[38]

In the seventeenth century, the British prophet of modern science Francis Bacon observed that "it is enough to check the growth of science, that efforts and labors in this field go unrewarded."[39] At the dawn of the nuclear age, SSRC's social science representatives wanted their enterprise to grow, and they concluded, after considering the risks involved, that growth would require rewards in the form of public support and recognition.

The Problems of Scientific Credentials and Social Relevance in the Nuclear Age

After the social sciences' ominous exclusion from *SEF* and associated conservative science policy proposals, their incorporation into Kilgore's liberal science legislation, and SSRC's decision to advocate their inclusion in the proposed NSF, legislative hearings in the fall of 1945 provided the occasion for more extensive consideration involving a wider range of participants. Representatives from the natural sciences, engineering, medicine, business, labor, Congress, and the executive branch commented on the competing proposals for a new science agency. Many individuals talked at least briefly about the social sciences. In this context, questions about the social sciences' scientific status and social relevance set off political fireworks and also underscored the power of conservative opposition within the scientific community.

Upon first glance the social sciences seem to have fared well at these hearings. Of the ninety-nine witnesses, not even one expressed outright opposition to federal funding for them. Of the forty-five witnesses who commented directly, thirty-seven favored their inclusion without qualification. A much smaller number approved their inclusion with reservations, while a few suggested putting them in a separate agency.[40]

However, Vannevar Bush maintained a cautious position that implied keeping the social sciences on the sidelines. Although he did not oppose them directly, he suggested the "foundation should allow an opportunity for effective integration and partnership between the natural and social sciences." He added that such a partnership "should be the result of careful study by the foundation after its establishment." Thus, as he did in *SEF*, Bush proposed the natural sciences deserved consideration first and foremost.[41]

A group of influential conservative scientists and science administrators who supported Bush's plans expressed a more antagonistic position. In *SEF* Bush drew special attention to the value of "basic science," to the vital importance

FIGURE 1. Vannevar Bush meets with physicists to discuss the cyclotron. University of California, Berkeley, Radiation Laboratory, March 1940. *Left to right*: Ernest O. Lawrence, Arthur H. Compton, Vannevar Bush, James B. Conant, Karl T. Compton, and Alfred Loomis. *Photo courtesy of Lawrence Berkeley National Laboratory.*

of research guided by purely "scientific" concerns "without thought of practical ends." Though in this definition basic science did not have a utilitarian orientation, Bush then explained that basic science served as the "pacemaker of technological progress," as the bomb and other wartime breakthroughs suggested. In addition, invoking a sharp distinction between the scientific and political spheres, Bush argued that scientific institutions and research should be insulated from political pressures and social interests.[42] Did these points about "science" extend to the social sciences? Not according to many of his allies who had ties to the defunct SAB, to the prestigious NAS, to the wartime OSRD, and to Bush himself.

Subject matter presented one problem. Researchers had not found any fundamental social laws comparable to those in the natural sciences, stated Bradley Dewey. According to this NAS member, president of the American Chemical Society, and MIT trustee (Bush's former academic home), social research therefore qualified as part of the humanities. In fact, the American Chemical Society as a whole opposed the inclusion of the social sciences, declaring them to lie "in the field of the humanities." Similarly, the engineer Boris A. Bakhmeteff explained that "immutable laws of nature" existed only in natural science, whereas social science dealt with "changing relations between men." Using an unforgettable analogy, Dewey concluded that "just as hair and butter should be kept apart," the social and natural sciences should not be "dealt with in the same legislation."[43]

Social science methods of study were different as well, stated the accomplished experimental physicist Karl Compton, who reasoned the proposed agency would therefore be more effective without the social sciences. Compton's views reflected his professional trajectory as a longtime NAS member, a former president of the American Physical Society, the main architect of SAB's unsuccessful Depression Era plans for publicly supported science, MIT's president for more than two decades, and a key OSRD member under Bush. Another physicist, Isadore I. Rabi, provided more specific methodological comments than Compton. Also a NAS member and an associate director at MIT's famed Rad Lab, Rabi appealed to the common view that natural scientists, especially physical scientists, uncovered universal truths about the world through experimentation of an unbiased, "quite objective" sort. In contrast, as seen by this Nobel laureate and prominent postwar federal science policy adviser, the social sciences had tremendous difficulty proving their claims in a rigorous, objective manner.[44]

Objectivity in the sense of political impartiality also concerned critics in Bush's circle who associated the social sciences with left-leaning social reform. The Johns Hopkins University president Isaiah Bowman stood out as the leading American geographer of the interwar era and had recently served as president of the American Association for the Advancement of Science. One may have expected Bowman to be sympathetic to the social sciences' cause because his scientific discipline included both social and physical studies, and his own pioneering works on regional geography tried to integrate the two. But Bowman always had greater confidence in the scientific validity of physical geography. Looking at the cultural and social sides of his discipline, Bowman found "horrible examples" of "confused thinking" and "shocking inaccuracy and superficiality." Being a close professional colleague of Vannevar Bush, a former NAS vice president, and the former SAB director, Bowman probably felt no obligation to side with his social science colleagues. According to his biographer Neil Smith, as Bowman "soured on Roosevelt, the New Deal, and social liberalism, his political conviction eclipsed even his passion for an expansive geography, and social science was the victim." In damning testimony Bowman told Congress "it is well-known that so much of human prejudice . . . and social philosophy enter into the study of social phenomena." In private correspondence, he added that in light of Senator Kilgore's connection to "labor," his social science–friendly bill might permit "doctrinal guidance" of social research supported by "millions" of public dollars.[45]

A related problem concerned the use of publicly funded social inquiry to support social agendas: if, as Bowman suggested, values could influence social research, this research could, in turn, promote those same values and associated policies. Federal support might thus strengthen preconceived viewpoints on political questions, warned the physicist Rabi. In a similar manner, Morris Fishbein, the conservative editor of the American Medical Association's main journal, an opponent of socialized medicine (i.e., national health insurance), and an advocate of Bush's science policy plans, asserted that letting in the social sciences would create a "great danger," because various interests could then manipulate the proposed agency for "political purposes." On a congruent note, the engineer Bakhmeteff cautioned that including social science could attract the interest of "pressure groups," making it impossible to keep NSF's administration "outside of any political influence . . . in a climate that is purely scientific." Henry S. Simms, also from the medical field, stated that if the agency welcomed the social sciences, the government might find itself supporting "crackpot schemes for altering the form of government."[46]

Thus, in promoting Bush's science policy plans, these scientists criticized the social sciences by appealing to key principles advanced in those plans. Certain points, such as Bradley Dewey's comment about the absence of laws in the social sciences, did not necessarily reflect any political or ideological position, but other points did, such as Simms's comment linking social science to crackpot schemes. These scientists also worried that the social shaping and social uses of the social sciences would compromise their insistence that massive public funding of science through a new agency should leave science free from political control and social interests. For Bush's allies, the social sciences' alleged contamination by politics and social values, as evidenced by social scientists' involvement with planning or labor, disqualified them as science, and, furthermore, such contamination threatened to provoke destructive partisan controversy. Thus their inclusion could jeopardize the creation of a central science agency governed by scientific interests as Bush specified and not political interests as Kilgore's liberal plans would permit.

Given this combination of political objectives, ideological concerns, professional interests, suspicions about the social sciences, and related philosophical ideals about the vital separation between politics and science, this group of Bush's allies surmised that an alliance with the social sciences would be a serious mistake. Summing up the general sentiment, Rabi asserted it would be unwise to allow the social and natural sciences "to sink or swim together."[47]

How did the social sciences respond to this multilevel assault? An invitation from Senator Kilgore gave the SSRC, "by virtue of its representation of all of the diverse social science groups," the opportunity to present its viewpoint during the 1945 fall hearings. While it is difficult to know how social scientists at large felt about SSRC's role here, a resolution from the American Psychological Association board of directors asked the council "to take whatever action is feasible to insure the inclusion of the social sciences," and an American Sociological Society (renamed the American Sociological Association in 1959) resolution claimed that "in the interests of the national welfare and safety" the social sciences needed proper recognition, support, and representation. Over the next year other disciplinary societies conveyed similar sentiments.[48] The council would therefore represent American social science with the tacit support of these societies. Council deliberations, as noted before, had already produced a consensus about the need to seek inclusion. But the council's scholars now had the difficult task of explaining to Congress why and how the social sciences should be included.

The Unity-of-Science Defense

At the fall hearings the SSRC's representatives included the economists Wesley Mitchell and Edwin Nourse, the sociologist William Ogburn, the psychologist Robert Yerkes, and the political scientist John Gaus. In their testimony these scholars focused on many issues that troubled their critics, but, not surprisingly, they advanced the opposite viewpoint. In brief, the council's strategy for winning inclusion presented the social sciences from a scientistic perspective, based on the premise that there was, in fact, a unity among the natural and social sciences. This position entailed, among other things, that social scientists should make a clear distinction between their ostensibly value-neutral and nonpartisan knowledge-making activities as scientists and their social or political activities as citizens. In an engaging analysis of boundary work in this episode, the sociologist of science Thomas Gieryn has proposed that social scientists "seemed to be divided and somewhat in disarray."[49] Indeed, conversations among social scientists indicated some differences of opinion about the wisdom of the scientistic strategy. Furthermore, during the interwar era many of the council's scholars had been deeply involved in the debate over advocacy versus objectivity. However, as this section and additional analysis later in this chapter will show, the SSRC-orchestrated effort to win a place for social science in the proposed agency kept such disputes about the nature and purposes of this enterprise largely out of public view.

Looking at the council's selection of accomplished scholars to testify reveals its scientistic strategy clearly:

Wesley Mitchell. Mitchell was a former president of the American Economic Association, the longtime research director at the National Bureau of Economic Research (NBER), a former member of the social science–led NRPB, a former SSRC chairman, and the chairman of SSRC's Committee on Research and the Federal Government. At the NBER, Mitchell had spearheaded an ambitious program of empirical, quantitative studies that aimed to place economic knowledge on a par with physics or chemistry and, equally important, beyond partisan politics. A world leader in studies of the business cycle based on detailed statistical analysis, Mitchell anticipated that through a combination of theory and empirical research economists would one day produce "an economics worthy to be called a science."[50] He also presented the council's formal statement to Congress at the 1945 fall hearings.

William F. Ogburn. A Chicago sociologist, Ogburn was also a former SSRC chairman, the director of the President's Research Committee on Social Trends (1929–1934), a consultant for many other governmental committees including the NRPB, and a president of both the American Statistical Association and the American Sociological Society (ASS). Well known for his work on cultural lag and quantitative indicators of social change, Ogburn encouraged his scholarly peers to reject the goal of improving the world in order to concentrate on scientific sociology. He prodded sociologists to abandon "the literary scholarly tradition" and to focus instead on the development of testable hypotheses and their empirical verification. In his 1929 ASS presidential address, he declared "science is interested in one thing only, to wit, discovering new knowledge."[51]

Edwin Nourse. An agricultural economist, Nourse was a recent American Economic Association president, a longtime leader at the Brookings Institution, and, like Mitchell and Ogburn, a former SSRC chairman. In 1947 he also became the first chairman of the nation's new Council of Economic Advisers (CEA). Known among other things for cultivating a detached scholarly posture, Nourse even insisted that CEA economists should "preserve the non-political character of the agency," though whether this agency could ever really remain unaffected by the ever-present and often powerful political pressures that shaped national economic policy remained a disputed issue.[52]

Robert M. Yerkes. A former president of the American Psychological Association, a leading primatologist, and an emeritus professor at Yale University, Yerkes described himself to Congress as a psychobiologist. Ever since World War One, when he had been in charge of the army's massive mental testing program, Yerkes had been trying to convince natural scientists and governmental officials that psychologists were "working in the spirit, with the objectives and in principle, with the methodologies of the other physical and biological natural sciences." Also a NAS member and the chairman of two controversial NAS committees that dealt with the sensitive topics of sex and race, Yerkes probably knew as much as any social scientist about the suspicions of NAS-associated natural scientists and science administrators who now supported Bush's science policy plans.[53]

John M. Gaus. The 1945 president of the American Political Science Association, Gaus came from the University of Wisconsin but soon took a position at Harvard University. As a public administration specialist, Gaus had gained renown in a field that distinguished between the interest-laden, rough-and-tumble political realm and the supposedly objective, apolitical realm of administrative expertise.[54]

Revealing as well, the SSRC did not choose any scholar who represented a contrary position to testify. It is particularly striking that the council did not include a historian, even though for two decades the American Historical Association had been a participating member in the council and historians took part in internal council discussions about postwar federal science funding. Presumably, SSRC leaders understood that placing a historian before Congress in this episode would have been risky, given the general American tendency to classify history as part of the humanities.[55]

Just as striking, the SSRC did not select a single critic of scientism to testify, though a couple of council members had made forceful critiques of the scientistic persuasion during the previous decade. Most famously, in his classic 1939 book *Social Science for What?* the sociologist Robert Lynd urged social scientists to make their value premises explicit and to declare their sociopolitical commitments, for example their support for democracy as opposed to fascism. Though he did not name his Columbia University colleague Wesley Mitchell directly, Lynd, who participated in various council activities and at one point served as its executive secretary, attacked the sort of detached scholarly position Mitchell advocated. In addition, during the mid-1930s the sociologist Louis Wirth wrote a major internal report for the council in which he explicitly criticized the council's commitment to a close analogy between the social and natural sciences. Wirth, as we will see later, opposed scientism and a value-neutral, instrumentalist standpoint for reasons similar to those offered by Lynd and other well-known left-liberal scholars such as John Dewey and Charles Beard. But the council did not ask Lynd or Wirth to present their views to Congress.

Thus the group of social scientists selected by SSRC to testify emphasized, without any dissenting voices among them, that social research was not, as Bowman charged, subject to human prejudice and social philosophy. Nor was social science the handmaiden of political interests or pressure groups, as Fishbein and Bakhmeteff held. The social sciences had once been "greatly confused by the mixing in of values with the consideration of knowledge," admitted the sociologist Ogburn. But he added that this unfortunate state of affairs had been improving as these disciplines matured.[56]

Besides highlighting the social sciences' commitment to value-neutral and apolitical inquiry, Ogburn also tried to reassure critics that social science did not prescribe, as did certain areas in the humanities. Instead, he explained that the social scientist worked like the natural scientist who might make poison gas but then refrained, as the bounds of scientific expertise dictated, from stating whether the poison gas "shall be used for spraying fruit trees or for killing human beings"—an analogy that should have been alarming in light of then-recent revelations about the Nazi regime's use of poison gas to commit mass murder. In Ogburn's words, the scientist aimed "not to tell us what to do" but to furnish the information necessary for politicians and others to make informed decisions. "Science properly considered does not undertake to say what ought to be done," Mitchell concurred.[57]

Congress also heard from SSRC's contingent that disinterested, value-neutral social inquiry had its utilitarian payoff, just like basic natural-science research. As Gaus observed, some critics implied that social research, being little more than "the recording of partisan and prejudiced attitudes," could have "no fruitful result." But he rejected this view. If "properly trained research workers imbued with scientific detachment and integrity" received "adequate resources," they could produce results of "inestimable practical value," agreed Mitchell. All the sciences work in the same manner, added Nourse. Social science, like natural science, produces "fuller and more accurate knowledge" of the "materials" and "forces" in the world, thus enhancing human control needed for a "safer and more satisfying existence."[58]

In sum, beyond any small differences in their testimony, these social scientists all presented the same clear message that the proposed agency should strengthen, not weaken, the "inherent unity of science," as Nourse put it. The "formal divisions" among the physical, biological, and social sciences were "arbitrary," he explained. "Existing demarcations or barriers among the several sciences . . . are entirely artificial," agreed Yerkes. If one looks at "major research problems" concerning the "national interest," one sees that "the traditional lines of demarcation" among the sciences have "little meaning."[59]

Yet, however often these social scientists repeated the mantras of intellectual purity, political neutrality, practical utility, and scientific unity, they were already in a hole. The council's effort to place social science above or beyond politics, in a strictly scientific realm, had not succeeded. As previously discussed, Bowman, Compton, and other supporters of Bush's policy principles had little confidence in the scientific basis and social value of the social sciences. Given their worries about the potential for political control over science, they now had an additional cause for concern as well. During the 1945 fall hearings, SSRC's formal statement and its social science representatives all favored a strong agency director who would serve at the president's pleasure. Senator Kilgore and President Truman also supported this proposal, while Bush and his allies worried that it would open the door to undue political influence on the agency's scientific work.[60]

Partisan Conflict and Political Vulnerability

As the NSF debate continued in late 1945 and 1946, the major conservative and liberal protagonists in the scientific and political spheres came to agree that unless the social sciences remained at a safe distance from the natural sciences, they might, as Rabi had warned, sink together. The process through which this consensus arose underscores the growing power of conservative, anti–New Deal politics in deciding the place of the social sciences in the proposed NSF. Furthermore, after the two sides reached that consensus, the social sciences received meager support from the Truman White House. To appreciate the full power of Rabi's warning, we first need to consider wider partisan struggles that contributed to the demise of a viable left-liberal political alternative in postwar America and that strengthened an alliance between conservative politicians and conservative scientists.

In the early postwar years, a contingent of liberal-left scientists active in the national science policy debate and elsewhere advocated a progressive left politics of science, as Jessica Wang's work on American science and anticommunist politics has so nicely shown. These scientists, including a number of participants in the newly formed and moderately liberal Federation of Atomic Scientists (later known as the Federation of American Scientists), hoped to bring about an internationalist world order. With the intensification of hostilities between the American-led and Soviet-led camps, this group of scientists feared that military patronage for American science and associated security restrictions would remain extensive in the postwar years. To promote a peaceful world order, they favored more open and cooperative foreign policies that included the international control of atomic energy. On the domestic side, these scientists supported New Deal–style antimonopoly and planning measures, including a national full employment policy. Accordingly, this left-liberal group supported Senator Kilgore's efforts to establish a science agency that included the social sciences and that would be responsive to various social interests.[61]

Opposed to these left liberals stood a more conservative contingent within the nation's scientific elite. A few exceptions aside, this group looked more favorably on the postwar expansion of military science, urged adversarial policies toward the Soviet Union, and opposed domestic policy initiatives they associated with heavy-handed social planning and excessive governmental regulation of the economy. In the NSF debate, this conservative group included Bush's personal friends and close professional colleagues, such as Compton and Bowman, as well as supporters from engineering, medicine, and chemistry, including Bakhmeteff, Fishbein, and Dewey.

In this broader political and ideological context, the question of how to view the social sciences would remain a decidedly partisan issue, just what SSRC's social scientists who declared their enterprise to be apolitical and their research to be value neutral hoped to avoid. On one side, the Democratic White House indicated the social sciences should be included. A revised draft of Kilgore's bill included them. During the hearings, Kilgore also declared his personal support for them, while SSRC's social science witnesses all testified in favor of his bill. Other testimony revealed a larger base of left-liberal supporters, including Truman's secretary of commerce Henry A. Wallace. An agricultural scientist by training, Wallace welcomed social and economic planning grounded in social science expertise, embraced the New Deal, and supported the NRPB as well its proposal for a system of national social security including national health insurance. A few years later, Wallace's commitment to cooperation with the Soviet Union led to a bitter split with Truman and then a crushing defeat when he ran as the Progressive Party's presidential candidate in the 1948 national elections. On the other side, a conservative group of scientists established an alliance with conservative Democrats and Republicans while they attacked the social sciences and favored Senator Magnuson's bill. Magnuson himself indicated that since natural-science funding had much higher priority, the social sciences, whose consideration might well delay the passage of satisfactory legislation, should be left out.[62]

As the NSF debate continued, the partisan divide grew wider, with scientists splitting into opposing groups. Alarmed by the Truman administration's position, Isaiah Bowman led a group called the Committee Supporting the Bush Report. A November 1945 letter from Bowman's group to Truman, signed by Bakhmeteff and Dewey among others, specified that including the social sciences would be a "serious mistake" and proposed that their support should come from "a separate body."[63]

The Committee for a National Science Foundation, a counterpart to Bowman's group, formed the following month under the leadership of the Harvard astronomer Harlow Shapley and the Nobel-laureate nuclear chemist Harold C. Urey. Both Shapley and Urey participated in left-leaning organizations, and their new committee came out in favor of including the social sciences. The committee's members included SSRC's contingent of Mitchell, Ogburn, Yerkes, Nourse, and Gaus as well as other prominent social scientists such as Louis Wirth and Talcott Parsons.[64] With the Urey-Shapley group's approval, Senator Kilgore introduced a new bill, contra Magnuson's bill, for a science agency that included a social science division.

Yet Kilgore, who sought a bipartisan compromise on other policy issues, soon retreated on the social science question. In February, Kilgore and Magnuson introduced a new compromise bill that called for a social science division, but that same month Kilgore's subcommittee said the agency's social science activities should be limited "until adequate planning studies" were completed. The parent Committee on Military Affairs took a similarly cautious stance, recommending that, at least initially, NSF support for social research should be "limited to studies of the impact of scientific discoveries on the general welfare and studies required in connection with other projects supported by the Foundation." As Kilgore explained, this move to limit the scope of social science activities reflected an agreement struck between leading scientists and members of the two subcommittees.[65]

Meanwhile, conservative legislators continued to barrage New Deal programs and related initiatives inspired by liberal social science planners and reformers. Though conservatives had killed the social science–led NRPB, its spirit survived in other political projects. Conservatives naturally set out to destroy them as well. Thus a coalition of conservative southern Democrats and Republicans repeatedly barraged social science programs in the U.S. Department of Agriculture (USDA). Prior to World War Two, the USDA had been among the most important centers for federal science funding and research, including influential work in rural sociology and agricultural economics. Beginning in the 1930s, USDA's Bureau of Agricultural Economics (BAE) developed social and economic planning programs, including efforts to make planning more democratic by involving poor rural and urban groups. But in the early 1940s conservative congressional interests prevented the BAE from engaging in further planning work. Congress then cut funding for BAE's social survey research. In 1946, while ASS president Carl Taylor proclaimed that "national planning is not dead," Congress simply forbade the BEA from carrying out such research altogether, effectively "crippling the work of social scientists."[66]

Conservatives in Congress also opposed an initiative to make the government responsible for maintaining full employment. Studies by the NRPB undertaken at President Roosevelt's request had outlined the case for a full-employment initiative, while young Keynesian economists helped to refine and promote it. As they saw it, such an initiative would help prevent another economic depression and assure gainful work for every American who desired it. Truman's 1945 reconversion address supported this initiative as well. But by the time the 1946 Employment Act became law, conservatives who opposed governmental economic regulation had successfully removed the original bill's strongest measures. Consequently, the enabling legislation did not make the government responsible for achieving full employment.[67]

A third noteworthy target of conservative wrath concerned race relations. For some time, well-known liberal social scientists such as the anthropologist Franz Boas had criticized racial segregation and other forms of discrimination. During World War Two Gunnar Myrdal, a Swedish economist and Social Democratic member of parliament, gained fame as the main author of a monumental study of American race relations, *An American Dilemma: The Negro Problem and American Democracy*. Myrdal's work, sponsored by the Carnegie Corporation, helped establish a postwar liberal consensus about the need to eliminate American racism.[68] Social researchers also contributed to the U.S. Office of War Information's support for racial integration in certain wartime production facilities. As seen by outraged segregationists in Congress and elsewhere, such efforts to promote racial equality and integration either constituted subversion or indicated the existence of a more comprehensive anti-American plot. In the context of these and other partisan struggles previously discussed, popular calls to eliminate or restrict the influence of "professors" in government during the mid-1940s typically focused on social scientists, as one SSRC document noted.[69]

As for the NSF debate, the growing power of conservative interests crushed the social scientists' quest for direct inclusion. Having lost Kilgore's strong support, the social sciences now became increasingly vulnerable to hostile legislators supporting Bush's plans. In another round of Senate hearings held during the spring of 1946, Republicans provided the most damaging charges. Compared with earlier criticisms put forth by natural scientists, these charges had more specific political content, though basic points about social science's questionable scientific credentials and leftish social agenda remained the same.

Federal support for social research would entail "promoting all the health legislation . . . all the housing legislation . . . all the other matters which come in under the all-inclusive term of 'social sciences,'" thundered the Ohio Republican senator Robert Taft. A heavyweight among congressional conservatives, Taft ardently opposed the New Deal, fought against the NRPB, and helped emasculate the 1946 Employment Act. His best-known legislative achievement came a year later in the Taft-Hartley Act, which passed over Truman's veto and sharply curtailed the powers of organized labor. Taft insisted a new science agency "should not turn to social sciences," which he called little more than "politics." These disciplines did not belong in a bill designed to promote "pure science, the discovery of truth," agreed H. Alexander Smith, a New Jersey senator and by this point the main Republican advocate for an agency based on Bush's vision.[70]

Moving in for the kill, Smith offered an amendment that would, among other things, have eliminated support for the social disciplines. Although the Senate rejected his amendment, the Connecticut Republican Thomas C. Hart put forth a new one to delete them. "No agreement has been reached with reference to what social science really means. It may include philosophy, anthropology, all the racial questions, all kinds of economics, including political economics, literature, perhaps religion, and various kinds of ideology," declared Hart. Like Smith, Hart said he saw "no connection" between social research and basic research in the natural sciences. Despite some supportive words from Kilgore, the Senate voted favorably on Hart's amendment. Congress then approved the Magnuson-Kilgore bill with Hart's amendment included.[71]

That spring the social sciences took a nosedive in the House as well. With the encouragement of Bush and Bowman, Representative Mills introduced a new bill designed "to carry out the recommendations contained in the Bush report [*SEF*]" and, consequently, omitted the social sciences. Contact with these disciplines would invite "political, social and economical [*sic*] disputes that would seriously interfere with its [NSF's] main objectives of supporting scientific activities in the natural sciences," warned Mills. Support for social science could lead to "hare-brained studies about things not capable of objective study," chimed Bowman. Expounding on the dire implications for American science, the anti–New Deal Ohio Republican Clarence J. Brown offered this colorful warning: "If the impression becomes prevalent in the Congress that this legislation is to establish some sort of an organization in which there would be a lot of short-haired women and long-haired men messing into everybody's personal affairs and lives, inquiring whether they love their wives or do not love them and so forth, you [the scientists] are not going to get your legislation."[72]

When support for the compromise Magnuson-Kilgore bill crumbled that summer, the social science question remained unsettled, but this respite proved to be fleetingly brief. National elections in November gave Republicans control of both legislative branches, thereby strengthening the hand of a decidedly anti–New Deal, anticommunist group that included Richard Nixon and Joseph McCarthy. After Senate conservatives placed the science legislation in the hands of a subcommittee chaired by Senator Smith, he pushed through a bill that did not exclude the social sciences explicitly, but neither did it include them. The bill's "whole emphasis" focused on the natural sciences, emphasized Smith. Though the Arkansas senator J. William Fulbright asked his colleagues to support a new initiative to include a social science division, his effort proved futile, and predictably so since Fulbright was a liberal internationalist and Democrat. After the Senate passed Smith's bill by a wide margin in the summer of 1947, the House and Senate agreed on another compromise bill that did not include the social sciences.[73]

At this point, only the president might have been able to reverse this downward spiral, but Truman made no such effort. His White House predecessor, Franklin Roosevelt, had facilitated connections between social scientists and the New Deal and also planned to underscore their importance to the nation in his undelivered Jefferson Day address. Truman, however, tended to be unenthusiastic or downright skeptical about them, despite the supportive stance in his reconversion address. A telling example comes from Truman's frustrating experience with his first Council of Economic Advisers. After failing to obtain clear policy advice from the council, at the time under Edwin Nourse's avowedly apolitical leadership, Truman quipped that he wanted a "one-handed economist," meaning an economist who would not tell him on the one hand "A," but on the other hand "B."

Though not anti–social science per se, Truman had little patience for the qualified arguments and endless debates typical of scholarly discourse.[74]

Furthermore, in the case of the President's Scientific Research Board (PSRB), Truman signaled his willingness to abandon the social sciences just as Senator Kilgore had. Truman created the PSRB with a mandate to develop a liberal counterproposal to Bush's *SEF*. Thus he could have told PSRB's chairman John Steelman, himself an economist and labor relations expert, to include the social sciences. But Truman did not. Consequently, PSRB's 1947 Report merely acknowledged their importance while leaving an investigation of their needs to someone else. In effect, the Democratic White House had retreated to the position favored by Vannevar Bush.[75]

As the NSF debate dragged on, social scientists watched nervously from their by-now-familiar position on the sidelines. Though the Hart amendment to exclude them was not the final act in this story, that amendment made their political vulnerability painfully evident. Few people after 1946 and hardly anybody after 1947 paid attention to the social sciences. To wit, Congress never even invited social scientists back to testify. In the best of moments they received some encouragement, mainly from liberal Democrats. Yet the Democratic Party did not present a unified stand in this episode, as indicated by the roles of Magnuson and Mills. Most troubling, Kilgore and the Truman White House abandoned them. One year after putting forth an amendment to include them, Fulbright gave up as well, noting that some of his legislative peers still wondered whether an amendment supporting the social sciences meant supporting "socialism."[76] Meanwhile, an alliance of conservative politicians and natural scientists situated social inquiry outside the scientific realm and inside the presumably very different realms of philosophy, ideology, and politics.

In the midst of these developments, SSRC discussions pointed out—and with good reason—that social scientists faced a problem of "salesmanship." These scholars needed to work harder on "clarifying" their functions and "convincing the public of their value."[77] The council would therefore continue its efforts to address the interrelated matters of scientific status, social utility, national contributions, public image, and federal funding.

Mounting a Public Relations Campaign: Scientific Status, Public Image, and Salesmanship

Practical concerns about cultivating public support also continued to loom large in SSRC's efforts to establish a more favorable understanding of the social sciences among scientists, politicians, and the public. In documents prepared for public consumption, the council always emphasized the unity of the sciences, while it left aside calls for social inquiry that addressed moral issues and problems of social justice head-on. However, behind the scenes a battle over the wisdom of this strategy simmered. Consider the council's dealings first with Louis Wirth, who advanced a pointed critique of scientism from a left-liberal perspective, and then with Talcott Parsons, who supported the council's scientistic strategy through the development of a much-anticipated but ultimately unsatisfactory report.

At the beginning of 1946, the council asked Wirth to prepare a memorandum about the status of the social sciences in the proposed NSF. A highly regarded figure from the Chicago school of sociology, Wirth chaired the council's Committee on Organization for Research in the Social Sciences at that time. A year later he became ASS's first Jewish president. In 1950 he served as the first president of the new International Sociological Association as well. A pioneer in the ecological approach to urban studies, Wirth wrote widely on race relations, minority problems, social planning, and social theory.[78]

In a postwar essay titled "The Unfinished Business of American Democracy," Wirth argued that the nation should use its newfound power "wisely to heal the wounds of an ailing world and to build a peaceful and a better way of life for ourselves and for all." Recognizing that "external strength" depended on "internal unity," Wirth called on postwar America to realize the promise of "equality of opportunity for all, irrespective of race, creed, or origin." New Deal–style public programs and policies in basic areas from employment to housing, social security, health care, and education should "validate" national "ideals."[79]

As Wirth thought about the social scientist's role in the troubled age ahead, he returned to the critiques of scientism raised by left liberals since the 1930s. In his view, the ideal of the disinterested investigator who worked with neutral methods to produce value-free results had severe flaws. In his preface to the first English translation of Karl Mannheim's classic 1936 work on the sociology of knowledge, *Ideology and Utopia*, Wirth said Mannheim's distinctive contribution had shown that "thought becomes fully comprehensible only if it is viewed sociologically," which involves tracing the foundations of "social judgments to their specific interest-bound roots in society." According to David Kettler and Volker Meja, Wirth's "commitment to an empirical and cumulative sociology" led him away from "the formation of anything like a Mannheim 'school.'" That Wirth did not found a "school" of thought seems certain. However, Wirth did continue to discuss basic elements of Mannheim's critical approach to the sociology of knowledge favorably and proposed they had deep relevance for the social sciences in postwar America. In addition, Wirth served as editor of Gunnar Myrdal's *American Dilemma* (1944), the landmark study of American race relations in which Myrdal argued that "valuations . . . driven underground, hinder observation and inference." Thus, to become "truly objective," social scientists would have to make their "valuations explicit," argued Myrdal here and in many other writings.

He saw "no other device for excluding biases . . . than to face the valuations and to introduce them as explicitly stated . . . value premises." Wirth, much like Myrdal and other left-liberal scholars including the pragmatist educator and philosopher John Dewey, the political scientist and historian Charles Beard, and the sociologists Robert Lynd and Karl Mannheim, believed that by openly recognizing the value orientation of research, social science would become more honest, more realistic, and more valuable in guiding governmental initiatives needed to realize democratic principles of equality and justice for all.[80]

With this conviction, Wirth tried to redirect the controversy about the social sciences in the proposed NSF. In May of 1946, Wirth complained that the Kilgore-Magnuson compromise bill provided only a reactive and dangerously narrow role for these disciplines. As noted before, by that time Kilgore's concession to conservatives had led to the elimination of a social science division. According to the bill, the proposed agency would use the social sciences primarily to study the impact of scientific discovery on the public welfare. This limited approach, cautioned Wirth, would confine social research "to the relatively narrow field of predicting and analyzing the impact of technology upon social life." Convinced that postwar America required advances in "ethical" and "political" "wisdom," he feared the effort to achieve "neutrality," the movement to focus on "small scale problems," and the quest for "technical proficiency rather than comprehensive understanding" would lead social scientists down the path of "irrelevance." Wirth understood that the social sciences were "centrally concerned with the nature of the good life and the institutions that serve it." Elsewhere, in an essay on the "social responsibility of social science," he added that "no amount of aping . . . the natural sciences will be sufficient."[81]

If one focuses only on the public testimony of social scientists in the NSF debate, one could easily surmise that social scientists had arrived at a widespread consensus about the need to extricate social research from the influence of social values and to eliminate critical analysis informed by political philosophy and ethical reasoning. However, Wirth's views together with the positions articulated by well-known figures on the liberal left suggest a more complex interpretation. In fact, in the early postwar years Lynd, Beard, Dewey, Wirth, and Myrdal all continued to write as before. Though Mannheim died in 1947, his views remained an important subject of scholarly debate.[82] Neither had critiques of the value-neutral, apolitical, and scientistic stance advanced by conservative politicians, scientists, and other voices vanished, as the barbs from Senator Taft, Representative Brown, and Vannevar Bush's natural-science allies have suggested and as other chapters will further explore.

Furthermore, when internal SSRC discussions during the mid- to late 1940s continued to generate further disagreement over the role of values in the social sciences, the council sought clarification through university study

FIGURE 2. Louis Wirth participates in the roundtable broadcast *Higher Education for All*. University of Chicago, January 25, 1948. *Left to right*: T. R. McConnell, Louis Wirth, Earl J. McGrath. *Courtesy of Special Collections Research Center, University of Chicago Library.*

groups. At the University of Chicago, arguably the single most important university in the development of American social science during the first half of the twentieth century, Wirth became chair of a new, SSRC-supported "Social Sciences and Values" Subcommittee that included scholars from a wide range of disciplines. In February 1949, Wirth explained to the subcommittee members that they would be discussing "what the social sciences have to contribute to ethics and social policy and how in turn ethical and policy considerations affect the social sciences." Far from being new, these issues were "as old as social science" itself, Wirth pointed out. The pressure of recent events had also made them urgent issues for the current generation. Given the growing "ideological conflict" between the "capitalist-democratic and the communist-authoritarian" systems, Wirth raised the following question: Could and should the social sciences remain "neutral"?[83]

Distressingly, after ten seminar sessions held in the following months, the Chicago group had arrived at no consensus, as glimpsed through this brief exchange between two of its members, the physicist Leo Szilard and the economist Theodore Schultz. Szilard: "Science should deal with consequences only. Not with valuative problems. When we take political action we step beyond science."

Schultz: "That is the real difficulty. Things of importance step over your definition of science."[84]

Reporting to the council, Wirth noted that the Chicago scholars from various disciplines could agree on little else besides the point that determining how values and social research influence each other remained a complex and unresolved problem. More generally, the SSRC-sponsored Chicago seminars had failed to determine whether the "methods and procedures" of the natural sciences could be "applied to social phenomena . . . especially in view of the complications of observation, experimentation and generalization as we find them in social relations."[85]

Yet neither the critical position Wirth outlined in 1946 nor the unresolved issues revealed through the university seminars in 1949 appeared in SSRC's public statements and related writings by its social science members about the NSF debate. Adopting Wirth's position as the council's own would have exacerbated its critics' concerns about important differences between the social and natural sciences while also confirming conservative suspicions about the political character of American social science. Besides, Wesley Mitchell and other SSRC scholars had already testified that no such differences existed, at least none that should call into doubt the social sciences' rightful place under a global scientific umbrella. Just as Wirth noted that the nation's external strength depended on internal unity, the council's leadership followed a similar principle in the postwar science debate. As we have seen, as early as the fall 1945 hearings, the council had decided not to parade fundamental disagreements about the intellectual nature, social purpose, and national contributions of social science across the political stage. Remember, too, that by the first half of 1946, the fortunes of American social science in the NSF debate were heading toward rock bottom. Meanwhile, the influence of its conservative critics grew stronger.

Moreover, while Wirth expressed his concerns about scientism and related issues like value neutrality through university discussions, scholarly publications, and privately circulated reports and letters, the council stepped up its effort to address the problems of scientific status and image management on the public stage in ways social scientists of Wirth's ilk rejected. As one council document from the summer of 1946 explained, "the long-range task of enhancing the prestige of social science . . . involves strengthening operational ties with natural scientists on all levels," complemented by "publicity directed both to layman and specialists in other fields."[86] Toward this goal, the SSRC purchased and distributed free of charge five thousand copies of a pamphlet, written by the liberal *New York Times* science editor Waldemar Kaempffert, that favored strong federal patronage for American science, a new science agency, and "a closer union of the natural and social scientists."[87]

With high hopes, the council commissioned a major national statement on the social sciences from Talcott Parsons, the leader of Harvard's new interdisciplinary Department of Social Relations, a future ASS president, and a towering figure in postwar social science more widely. For this work, Parsons received $10,000, equivalent to double or even triple the average social scientist's annual salary at the time. Parsons also joined the council's newly reconstituted Committee on the Federal Government and Research, now chaired by the psychobiologist Robert Yerkes. "Several times" committee members emphasized the need to correct "the mistaken concepts of social science research held by many [natural] scientists," and they hoped Parsons's finished statement would have a large impact on public discussions, "analogous" to Vannevar Bush's already famous report on the natural sciences.[88]

Yet, at first glance, Parsons might seem no better suited than Wirth for such a sensitive task. Not only were the two sociologists on good professional and personal terms, but the younger Parsons also had great respect for Wirth.[89] As a scholar, Parsons became best known for his analysis of social systems and his attempt to develop a universally valid theory of social action. Though his interest in the stability of social systems later came under heavy fire for its conservative implications, Parsons was not politically conservative by the standards of his day. More of a liberal Democrat and much like Wirth in his political outlook, Parsons had strong sympathy for New Deal ideas regarding national economic regulation—though Parsons's economic views became more moderate as the postwar era unfolded, a transition skillfully explored by the intellectual historian Howard Brick. In addition, Parsons found the rise of the political right in postwar America alarming.[90] Both positions could easily antagonize conservative critics of the social sciences.

Parsons's intellectual orientation also embraced certain points about social science inquiry that some critics used to challenge the claim that the unity of science extended to the social sciences. Most notably, Parsons's grand project required attention to subjective, value-laden components of social life. Drawing on earlier European social theorists, Parsons took a keen interest in the interpretive sociology of Max Weber, who insisted that social inquiry could be scientific and objective but that it also needed to pay attention to subjective meanings, a task with no close parallel in natural-science inquiry. On a related point, Parsons urged social scientists to study social values in a given society in order to illuminate how they functioned to promote or undermine social order, another dimension of social inquiry absent from the natural sciences. In emphasizing the voluntaristic basis and teleological orientation of human action, Parsons also rejected the natural science–oriented behaviorism of psychologists such as John Watson.[91]

FIGURE 3. Talcott Parsons delivers a radio address. Michigan State College, 1955. *Courtesy of Michigan State University Archives and Historical Collections.*

Yet in other respects Parsons could seem a wise choice, for he firmly believed that in the big picture, the social and natural sciences were complementary parts of a unified intellectual and professional enterprise. He elaborated on the nature of this unity in a series of articles regarding the NSF debate published in 1946 and 1947. Concerned that "naive, popular misunderstandings" had portrayed the social sciences as "a haven for crack-brained reformers," "a glorified form of social work," and a hotbed of "sexual libertarianism" that could not "rise above partisan politics," Parsons claimed that, in fact, it was "impossible to draw any distinct line between the natural and the social sciences." He further proposed that if the natural sciences' practical accomplishments were more impressive, the nation should "act in terms of future promise" by stimulating the advance of social science knowledge and corresponding applications, which he called "social technology." Though far from new in the annals of American scholarship, this engineering perspective resonated especially well with social scientists' recent wartime contributions.[92]

Parsons also believed social scientists could and should establish fruitful relationships with natural scientists, at the level of ideas and institutions. He greatly admired the Harvard biochemist and polymath Lawrence J. Henderson, the central figure in a circle of Cambridge intellectuals interested in applying the concept of the social system to the social sciences. Parsons, as much as anyone, carried that effort into the future. In a major postwar theoretical study titled, simply, *The Social System*, he set out to examine the process of social interaction "as a system in the scientific sense."[93] In addition, Parsons served as chairman for the Cambridge Committee of Natural and Social Scientists, a group that addressed problems regarding atomic energy.[94] As well, he agreed with SSRC's position that exclusion from the proposed NSF "would impose a great handicap" on the social sciences.[95]

Thus, contrary to Wirth's position, Parsons urged social scientists to seek public support by "riding in on the coattails of the natural scientists." This curious phrase possessed an equally curious origin that highlighted Parsons's determination to work with conservative critics. The phrase originally came to Parsons from the pen of a hard-core conservative in Bush's camp, John Teeter, who had advised Parsons that regarding federal funding, "a coattail ride" would be "better than none." Parsons recognized some social scientists found the prospect of riding natural scientists' coattails "distasteful." Nevertheless, hoping social scientists would be permitted onto the natural science–oriented "endless frontier" as described in *SEF* and in subsequent legislation, Parsons incorporated Teeter's language into a published account of the NSF controversy.[96] In doing so, Parsons revealed an eagerness to reassure conservative critics in the scientific and political communities that the social sciences posed no threat to their goals and interests. After all, those who ride your coattails might be a drag, but they are hardly able to challenge your leadership.

In a lengthy draft of his SSRC-commissioned statement, the Harvard star addressed the major points of public dispute about the social sciences in a predictable fashion.[97] Parsons claimed that one could not say with reason that "the scientific method" should be used to study "nature" but not "man in society." Against the suggestion that the social sciences had a closer kinship with the humanities, he characterized the humanistic approach as "essentially different," oriented more toward "appreciation" than toward "analysis, prediction, and control." Unfortunately, however, misplaced associations between social science and social reform often confused public figures who needed to understand the capabilities and limits of the social sciences at this critical historical juncture. Regarding this important last point about limits, Parsons emphasized that modern social research had a highly technical character and thus lent itself only to piecemeal applications, not large-scale social reconstruction. Skeptics could thus rest assured that American social scientists did not advance dangerous philosophies of life calling for the "total reconstruction of society," such as "Marxism." On the matter of federal patronage, Parsons, echoing *SEF*, emphasized the need to support "pure research," free from the "nonscientific pressure" associated with applied studies.[98]

In sum, Parsons worried about the spread of "anti-longhair sentiment" in Congress and sought to counter the common and at that time typically conservative charge that social inquiry was political and ideological rather than strictly scientific in nature. To do so, he followed SSRC's public relations strategy of underplaying any differences between the social and natural sciences at the levels of methodology, epistemology, and practical utility. "The whole argument of this report is that social science logically belongs in the proposed agency," he concluded.[99]

But in stark contrast to *SEF*'s powerful impact, Parsons's manuscript encountered paralyzing problems. One suspects the presence of conflicting visions of social science made it difficult for Parsons to craft a widely acceptable public document. His personal enthusiasm for the "rational, 'engineering' control" of social relations and countercyclical economic policies would have infuriated anti-Keynesians and other conservative critics. A further problem concerned the fit, or lack thereof, between product and consumer, partly due to Parsons's turgid prose.[100] A number of Parsons's peers suggested his manuscript would be inaccessible to the intended nonspecialist audiences, namely, natural scientists, politicians, and the educated general public. Samuel Wilks, a prominent mathematical statistician and strong proponent of quantitative social analysis, explained to SSRC president Pendleton Herring that even after going over parts of Parsons's draft three times or more, he still found it difficult to read. Herring regretfully found himself in agreement with Wilks and others that the manuscript needed "drastic revision." While Parsons made some further changes over the next couple of years, he never completed his report for publication.[101]

By the time President Truman signed NSF's 1950 enabling act, then, the council's ongoing efforts to clarify the social sciences' scientific credentials and practical value for the relevant political, natural science, and public audiences had accomplished little. The ambitious SSRC-sponsored public relations project involving Talcott Parsons fizzled out, thereby dashing the hope of creating a social science counterpart to Bush's *SEF*. Meanwhile, the SSRC-sponsored university seminars on values in the social sciences, especially the Chicago seminar led by Louis Wirth, highlighted continuing disagreement about the fundamental issues at stake. Moreover, the more general problems that prompted SSRC's efforts in image making and salesmanship remained largely beyond social scientists' control in ways that kept them on the defensive and often on the margins of the main action. After all, during the late 1940s neither Wirth nor Parsons nor any other social scientist had a chance to testify before Congress in the ongoing NSF debate. In fact, except for the one day of hearings set aside for SSRC's representatives back in the fall of 1945, social scientists had little opportunity to shape the course of national science policy in this key episode.

As for the status of the social sciences in the NSF, the agency's 1950 enabling act did not mention them specifically. Thus the NSF had no clear mandate to support them, but a clause referring to the "other sciences" left open the possibility that the agency could do so.

Conclusion

This chapter has argued that the NSF debate became a pivotal episode in the early postwar history of American social science. The debate itself helped to make the federal government a much more important participant in the long-standing argument about scientific legitimacy, social utility, and political relevance. In this episode, politicians, natural scientists, social scientists, and representatives from other major sectors of American society had a unique opportunity to express their views on fundamental issues, such as the relationship between the social and natural sciences, the possibility and desirability of separating social science from value-laden inquiry, ideology, and politics, and the potential benefits and drawbacks of federal funding. As we will see in chapter 4, the new NSF would also become an important locus of discussions related to such issues.

The NSF debate made visible certain features of the postwar political and scientific contexts that had major implications for American social scientists as well. These features included the powerful presence in the emerging postwar federal science establishment of natural scientists, especially physical scientists, and among them especially conservative physical scientists. For a combination of reasons including their political objectives, their professional interests, their commitment to a sharp distinction between politics and science, and their skepticism about the scientific character and political nature of the social sciences, a group of influential conservative scientists associated with Vannevar Bush's position criticized the proposal to include the social sciences. Partisan politics in this episode revealed a deep pool of hostility toward the social sciences among conservative legislators as well, some of whom formed an alliance with conservative scientists. Powerful figures such as Senator Robert Taft criticized the involvement of the social sciences with New Deal liberalism, suggested they had an affinity for socialism, and claimed they belonged in the realms of politics and ideology rather than science. Furthermore, the course of the NSF debate indicated that support for the social sciences from their predominantly liberal allies had only a modest foundation, as seen most clearly when Senator Kilgore, seeking a bipartisan compromise, abandoned the initiative concerning a social science division and when President Truman retreated from his earlier declaration that the social sciences should be included.

As for the efforts of social scientists themselves, discussions at the SSRC and the council's Committee on the Federal Government led by Wesley Mitchell initially highlighted the danger that extensive federal funding could lead to a dangerously harmful level of political control over American social science. Though this concern never disappeared entirely from view, council discussions quickly shifted direction by the time of the 1945 fall hearings, as social scientists became worried that exclusion from the proposed NSF would have harmful consequences, and as council leaders became worried about alienating influential natural scientists.

Under these circumstances, the SSRC led an effort to convince the social sciences' many critics that they belonged in a unified scientific enterprise led by the natural sciences. The sociologist George Lundberg, an ardent advocate of the unity of the sciences, scientism, and value neutrality, summed up the thinking behind this SSRC-led effort well in his opening quote to this chapter: "If social scientists aspire to the status and position and public estimation of other scientists, they must subject themselves to standards of the kind recognized by other scientists and by the public. That is, they must specify criteria that distinguish social scientists from that vast array of camp followers, reformers, propagandists, and social workers, which today dominate even most of the professional organizations of social scientists." While seeking inclusion on the natural science–oriented "endless frontier," the council's social scientists also indicated their commitment to work on an ostensibly apolitical, value-neutral, and hence "endless" frontier. In doing so, they indicated their willingness to leave aside the sort of direct engagement with human values and advocacy of social ends called for by Louis Wirth and other prominent left-liberal scholars including Lynd, Beard, Dewey, Mannheim, and Myrdal. Of course, SSRC's representatives had a strong motivation for taking this position, as a number of their conservative critics attacked their support for a wide range of social issues—racial equality, housing legislation, sexual freedom, labor unions, et cetera—and claimed their involvement with such issues rendered their work less than scientific and more akin to a political project.

Though my analysis has emphasized that the council's social scientists acted strategically and selectively, I do not mean to suggest that Mitchell, Nourse, Ogburn, Yerkes, Gaus, and Parsons were misrepresenting their own views. No doubt, many of them and many other social scientists as well claimed in numerous contexts beyond the NSF debate that social and natural science had a lot in common at the levels of epistemology and methodology and, more specifically, that social science knowledge could and should be value neutral and apolitical. However, Parsons himself sometimes recognized important differences in subject matter and research methods. Yet he chose to ignore or minimize the relevance of such differences in his writings about the NSF debate and in his stillborn essay for the SSRC. In this respect, he contributed to SSRC's wider effort to maintain a united front by keeping criticisms of scientism out of public view.

More important for the broader history of American social science, it is not obvious that those critiques of value-neutral inquiry, a disinterested professional stance, and a technical, piecemeal approach to solving social problems had been pushed aside by SSRC's scholars for convincing reasons. During the ferment of the 1930s, the wisdom of the scientistic project had inspired heated debate. Though wartime conditions did not allow much time for concentrated attention to the relevant issues, they reappeared in scholarly discussions after the war. Wirth's writings offered an articulate and thoughtful analysis about important dissimilarities between the social and natural sciences and about the need for ethical wisdom in the precarious nuclear age. In addition, when one probes beneath the statements prepared by the SSRC and its scholars for public consumption, one finds a continuing controversy about the role of values in social science and about the social responsibilities of the social scientist. Of course the role of natural science and political critics in this episode also reveals that claims about the apolitical and value-neutral character of the social sciences remained unconvincing to many conservative figures.

In assessing the upshot of this episode, Thomas Gieryn has proposed that "all's well that ends well." He reasons that despite the difficulties encountered by the SSRC, its elite contingent of social scientists, and their supporters, the NSF did begin funding basic social research, starting in the early 1950s.[102] Indeed, as we will see in the following chapters, social scientists obtained significant support not only from this new agency but also from other powerful patrons. However, social science funding also continued to pale in comparison with the quantities allocated to the natural sciences. In 1951, the sociologists Robert Merton and Daniel Lerner claimed with dismay that relative to the level of support available to natural scientists, funding for social researchers was "small and pitifully inadequate to their needs."[103] Moreover, for American social scientists the effort to secure funding remained linked to controversies about their disciplines' scientific foundations, social utility, and political relevance. The following chapters explore these issues by examining the development of the postwar funding system and social scientists' efforts to negotiate the basis for their support from major new patrons. Chapter 2 focuses on their struggles to secure funding within the enormous and powerful defense science establishment.

2

Defense and Offense in the Military Science Establishment

Toward a Technology of Human Behavior

American scientists are still struggling to reconcile their eighteenth-century devotion to science as a system of objective and dispassionate search for knowledge and as a means for furthering the welfare of mankind in general, with the twentieth-century necessity of using science as a means for strengthening the military power of the United States.

–Don Price, political scientist and consultant to the military science establishment, 1954

The basic general requirement is for a technology of human behavior suited to assist the "managers" of military effort in decision making about people.

–Charles Bray, military psychologist, 1960

By the end of World War Two, prominent figures from many sectors of society recognized that the future of American science would depend greatly on the growth and character of the postwar federal science establishment. And by midcentury what seemed like a likely scenario at the war's end had become a powerful reality: "The mighty edifice of government science dominated the scene in the middle of the twentieth-century as a Gothic cathedral dominated a thirteenth-century landscape," wrote the historian of science A. Hunter Dupree.[1] Whereas federal research and development (R&D) expenditures had amounted to less that $100 million in 1940, a dozen years later the federal contribution was nearly twenty times greater, almost $1.9 billion.[2] Looking only at academic science funding, federal dollars comprised over 50 percent of the nation's total contribution in the early 1950s, over 60 percent by the decade's end, and nearly 75 percent by the mid-1960s.[3]

Building on wartime precedents, American science also remained heavily involved with the military and other national security agencies. In 1935 the military portion of federal science expenditures had been only 25 percent. But by midcentury the DOD administered 53 percent of all federal funds for American science, and the Atomic Energy Commission, a civilian-controlled agency that supported a great deal of defense-related research, administered 36 percent.[4] Thus together these two agencies provided nearly 90 percent of all federal science funding, a stunning development that has led some concerned voices ever since then to worry about the militarization of American science. Also building on the wartime pattern, the largest science agencies in the postwar years focused predominantly on promoting work in the physical and engineering sciences needed to develop atomic bombs, conventional weapons, and weapons systems. But the military, other national security agencies, and other parts of the burgeoning federal science system also supported a breathtaking range of research throughout the natural sciences, for example in cryptography, geology, geography, medicine, meteorology, oceanography, and seismography.

Though the transition from hot war to the Cold War did not happen overnight, that transition took place quickly enough. By midcentury the major sectors of American science were mobilized for the indefinite future. Thus in 1954 the political scientist and DOD consultant Don Price could write, without singling out any field of research in particular, that "American scientists" needed to accommodate themselves to the "necessity" of using their work to strengthen the nation's military power.

This chapter focuses on the development of military patronage during the first two Cold War decades and examines the importance of military funding for the ongoing debates about the social sciences' scientific identity, practical uses, and political relevance. From the late 1940s through the early 1960s the social and psychological sciences obtained extensive support from a number of research organizations designed to serve the nation's defense needs, including the Defense Department's Research and Development Board (f. 1947) and a number of other research units established between the mid-1940s and mid-1950s to serve the more specific interests of the army, the navy, and the air force. As the sources of defense funding proliferated, the overall level of support for these sciences rose significantly as well.

In the process, the military's role in the emerging Cold War patronage system became enormously important by providing strong encouragement to social and psychological research pursued within a scientistic framework, as my analysis will show. Scholars who worked with military agencies during

World War Two and then during the Cold War years consistently claimed that by promoting research that met universal standards of scientific inquiry, the military patron stimulated the social sciences in exactly the right directions. These advocates argued, in addition, that the results of such research would fulfill the nation's practical military and national security needs in two powerful ways: by placing decision making about military operations and strategic affairs on a rational basis, and by developing effective means to predict and control human psychology, social systems, and political affairs in a manner that would advance the nation's Cold War objectives. These views then informed the development of scholarly research fields that received extensive military support and had significant policy influence. This chapter will consider three such fields: communications studies; the decision sciences, including operations research, systems analysis, and game theory; and the science of strategy, including work that initially focused on the threat of nuclear war but by the late 1950s and 1960s also supported counterinsurgency research.

Yet, as I will also explain, efforts to strengthen the military–social science partnership along the above lines also encountered a number of criticisms and difficulties, which makes the story of military patronage more complex than it appears when one focuses on research fields that attracted substantial funding, which historical scholarship has typically done. The previous chapter has already suggested that the difficulties faced by social scientists during the postwar NSF debate had implications that went far beyond that particular episode. During World War Two and then during the early Cold War years, conservative political criticisms extended to national security–oriented agencies that became involved with the social sciences. Meanwhile, from a different political and ideological standpoint, some liberal scholars proposed that military-funded work encouraged intellectually impoverished, ethically questionable, and antidemocratic tendencies in the social sciences. Within the defense science establishment social scientists also struggled with their subordinate position vis-à-vis influential natural scientists who continued to raise serious doubts about the social sciences' scientific status, an important point that emerges most clearly through an examination of unpublished documents, private correspondence, and behind-the-scenes negotiations. In addition, in the late 1950s and early 1960s, probing analysis from the nascent New Left suggested that at defense-funded research centers, the boundary separating allegedly objective and value-neutral scientific inquiry from the political assumptions and militaristic orientation of American Cold War strategic thinking seemed to have vanished.

Social Science War Stories during the Awkward Transition

As World War Two came to an end, the future of military funding for the social and psychological sciences seemed rather uncertain.[5] For many wartime social research projects and their sponsoring agencies, the war "ended less with a bang than a whimper," in the apt words of Barry Katz.[6] Consider the situation in psychology. As of late 1943, more than 1,000 of the nation's 4,500 psychologists had been involved in military work concerning "some of the very practical problems of winning a war," explained the psychologist Charles Bray. Psychologists had contributed to military recruitment, screening, and training programs. They had studied ways to make military leadership more effective and to strengthen social cohesion among military units. They had also contributed to psychological warfare efforts. During the Cold War, the military would become the largest patron of psychology. But, in the early postwar years, whether the military would continue its extensive engagement with psychology seemed unclear. Writing in 1948, Bray noted that "many psychologists expect that military psychology will die away in the days to come, as it died away in the days following World War I."[7]

The first part of this section suggests that the key developments in American partisan politics and national science policy that had placed the social and psychological sciences on the sidelines during the NSF debate also posed difficulties for scholars interested in working with national security–oriented agencies during the early postwar years. Some liberal scholars also raised serious doubts about the intellectual value and political implications of military-funded work. Yet, as an analysis of social science war stories in the second part of this section reveals, social scientists and their supporters more commonly presented an optimistic picture of military-funded research. This perspective, which had much in common with the scientistic viewpoint put forth by SSRC scholars during the NSF debate, would also become deeply influential in the evolution of the military–social science partnership through the mid-1960s.

During the second half of the 1940s and continuing into the 1950s, swelling anticommunist sentiment and associated political developments became pervasive issues for natural and social scientists. In response to escalating international tensions and hostile partisan attacks, the Truman White House launched a massive program to root out subversive elements in the federal government. Subsequently, the FBI carried out loyalty checks on some two million federal employees. In addition, the U.S. attorney general developed a list of organizations suspected of engaging in un-American activities. And by using the 1940 Smith Act, which enabled the government to prosecute anyone who promoted communism, the Truman administration crushed the American Communist Party's leadership. Under these circumstances, security clearances and loyalty investigations became a normal part of life for American scientists and scientific institutions. Because of their critical importance to secret weapons programs, physical scientists came under heavy investigation by the FBI, the House Committee on Un-American Activities (HUAC), and other pillars of the antisubversive campaign.

If, as Jessica Wang has convincingly argued, physical scientists had "no place to hide" by the early 1950s, social scientists faced similar problems.[8] The brunt of antisubversive attacks fell on scholars who at some point had expressed leftist sympathies. However, the politics of anticommunism also focused critical attention on liberal social scientists who promoted what they considered to be progressive viewpoints, such as advocating racial equality, challenging the national-security state where it seemed to be subverting rather than protecting freedom, and proposing that government should undertake measures to ensure a more equal distribution of wealth and greater social security. Liberal scholars who took up these issues often directly opposed communism as well. Nevertheless, many of them became vulnerable to anticommunist scrutiny, including such important midcentury figures as the sociologist Talcott Parsons and the anthropologist Ashley Montagu.[9]

The anticommunist campaigns together with strong conservative political pressures naturally created difficulties for federal agencies closely involved with the social sciences—and also for the large private foundations and the new NSF, as explained in the next two chapters. Toward the end of World War Two and continuing during the postwar Red Scare, a number of agencies, projects, and policy initiatives associated with liberal social scientists suffered sharp criticism, steep reductions in their funding and scope, and sometimes complete elimination. The previous chapter discussed a series of cases involving the National Resources Planning Board, the Department of Agriculture, and the Full Employment Act.

Additional high-profile cases that raised concerns about left-leaning social researchers involved agencies working directly or indirectly on national security matters, including the Office of Strategic Services (OSS). Created by a presidential military order in 1942, the OSS developed strategic information and carried out "special operations" in the areas of psychological warfare, sabotage, and other guerrilla activities. At its peak, the agency employed about 13,000 people. Its Research and Analysis Branch, which alone employed nearly 1,000 personnel, included social scientists from all the major disciplines, many New Deal supporters, and a few more radical figures such as the Marxist economist Paul Sweezy and the Frankfurt school scholar Herbert Marcuse. During most of the war, OSS director William "Wild Bill" Donovan benefited from the support of President Roosevelt. But following Roosevelt's death, Donovan and the new president, Harry Truman, did not get along well. Taking advantage of this rift, FBI director J. Edgar Hoover, Donovan's longtime nemesis, told the press that

Donovan planned to turn the postwar OSS into an American gestapo. Conservative critics also encouraged the agency's closure, though its legacy lived on in its famous successor agency, the CIA. Over half a century later, some conservative voices still claimed that the OSS had been infiltrated by communists.[10]

The Office of War Information (OWI), another agency that was created in 1942 and that relied heavily on social research to fulfill its responsibilities, also suffered from right-wing criticisms. OWI personnel included many New Deal supporters, some with Communist Party ties. The economist James Warburg, who served as the deputy director of the agency's overseas propaganda activities, provided personal financial advice to President Roosevelt as well. Not surprisingly, congressional conservatives viewed the OWI as a Democratic tool and, even worse, a hotbed of subversion. As a 1944 *Time* magazine article reported, these critics accused OWI's "psychological warriors" of being "leftish." Moreover, conservative attacks once again took their toll, first by supporting sharp restrictions on the scope and resources of the agency's domestic activities, and then by securing its closure altogether in September of 1945.[11]

After President Truman moved the Voice of America (VOA) from the defunct OWI to the State Department, conservative critics resumed their assault. During the early Cold War years, the VOA served as "the nation's ideological arm of anticommunism," as explained by the historian David Krugler. The agency thus became responsible for disseminating information and propaganda in the battle against the Soviet Union and, more generally, for promoting American viewpoints around the world. Though conservatives could agree in general with VOA's anticommunist agenda, some of them objected to the program's close ties to liberal Democrats and questioned the value of its propaganda activities. In 1947 congressional conservatives succeeded in slashing VOA's budgetary appropriations. They also blocked legislation to make it a permanent agency.[12] A few years later, Senator McCarthy led a congressional investigation of the VOA.[13] Persistent pressure resulted in the resignation of top officials, project cancellations, and yet another reorganization, as the Eisenhower administration moved the embattled VOA from the State Department to the new U.S. Information Agency.

In addition to the difficulties noted above, social scientists found themselves relegated to the sidelines within the enormous defense-oriented Cold War science establishment, and not only during the immediate postwar years, as indicated in the previous chapter, but for the indefinite future as well. From the mid-1940s to mid-1960s an elite group of physical scientists served as the scientific community's main representatives in federal science circles and military programs, thereby helping to crystallize what Paul Hoch has called the "military-industrial-scientific alliance."[14] Most of these scientists had held important positions in the major wartime nodes of scientific research, including the Manhattan Project, MIT's Radiation Lab, and the OSRD. After the war, they served on high-level advisory committees that oversaw the development of defense research programs, that coordinated scientific activities carried out at military, academic, and industrial sites, and that offered advice on national security matters. Among the nation's science policy elite during these years, the near absence of social scientists is remarkable.

Remarkable, yes, but not surprising, because many physical scientists still suspected that social scientists dealt "with problems that are full of value judgments" and thus were "handicapped in adopting a truly scientific attitude," as DOD consultant Don Price noted in 1954.[15] Indeed, a number of influential physical scientists from the Manhattan Project generation continued to question the scientific standing of the social sciences, as they had during the postwar NSF debate. In a paper originally presented at the 1948 annual meetings of the American Association for the Advancement of Science, Samuel Stouffer, the Harvard sociologist and former research director of the massive *American Soldier* wartime study, described this difficulty in colorful terms. As Stouffer put it, the scientifically minded social scientist during those years often looked into the mirror "wistfully," "like a fourteen-year-old boy" hoping that the "faint down on the upper lip gives promise of turning into whiskers." However, his "kindly elders" from the "fields of physics and chemistry" tended to regard him "with a certain tolerant amusement." "No son, those ain't whiskers yet," the "wise scientists" regularly concluded, reflecting their belief that the social sciences had not yet progressed beyond the "adolescent" stage.[16] Throughout the 1940s, 1950s, and 1960s, expressions of this skeptical attitude continued to come most often from the conservative wing of the scientific community. As noted in the introduction, Edward Teller, "father" of the H-bomb, told an academic audience that the social sciences had no more scientific standing than Christian Science.

Meanwhile, some scholars raised hard questions about the value of military-supported research itself. During the early 1940s, wartime pressures had typically required researchers to embrace a practical orientation. According to the psychologist Dorwin Cartwright, by the end of 1942 "virtually all research activities of social psychologists were oriented toward technological problems rather than strictly scientific ones." Another psychologist, Fred Sheffield, agreed: "everything was urgent, especially after Pearl Harbor. . . . None of us was thinking about [contributions to social science]. . . . We were doing applied research . . . for the sole purpose of getting answers useful to the War Department."[17] In short, as if to answer Robert Lynd's query "Knowledge for What?"—the title of his landmark book published on the eve of World War Two—wartime conditions had focused social scientists' attention on specific tasks handed to them by their superiors.[18] Given this limitation, the relevance of wartime research could thus seem minimal in the postwar years, and even worrisome from an ethical perspective.

In a 1949 critique of the *American Soldier* study, Lynd himself warned scholars that their expertise was "being used with great skill to sort out and to control men for purposes not of their own willing." Seduced by the funding offered by the U.S. Army, social researchers had apparently accepted the managerial problem of turning "frightened draftees into tough soldiers who will fight a war whose purposes they do not understand." Lynd found the manipulative implications of military funding disturbing and dangerous: "With such socially extraneous purposes controlling the use of social science, each advance in its use tends to make it an instrument of mass control, and thereby a further threat to democracy."[19]

Other criticisms of that project focused more directly on its intellectual limitations, especially its scientistic orientation. According to the sociologist Nathan Glazer, the *American Soldier* study revealed an "overpowering obsession" with a model of inquiry taken from the "physical sciences," which, in his view, consequently led to advances only in "techniques" but not in "understanding."[20] Similarly, the prominent historian Arthur Schlesinger, Jr., asserted that in this study one found a sociology that "whored after the natural sciences." As he saw it, the resulting cold, wooden, and altogether inadequate analysis presented the American soldier as he existed "neither in life nor in history" but only in "some dreary statistical vacuum."[21]

Following the acerbic commentaries from Lynd, Schlesinger, and Glazer, one might have worried about the larger implications of any future partnership between the social sciences and the military. And we have already seen that, during the postwar NSF debate, Wesley Mitchell's SSRC Committee and the University of Chicago's Louis Wirth raised the more general concern that federal funding might push social research in a technocratic and antidemocratic direction, with Wirth's position also noteworthy for its biting critique of scientism. Yet only a small minority of critical liberal voices expressed these views with any force during the mid- to late 1940s, while a much larger number of social scientists and their supporters argued just the opposite.

In a wave of articles and books, this larger and more influential group claimed that wartime social science work had many virtues and could thus provide a solid basis for future progress. The anthropologist Alexander H. Leighton told two stories, one about a group under Leighton's leadership that had been responsible for helping administrators keep order at the Japanese Relocation Camp in Poston, Arizona, and the other about a group that had carried out the postwar Strategic Bombing Survey in Japan. Daniel Lerner, a sociologist, related the history of psychological warfare operations directed against Nazi Germany. The psychologist Karl Dallenbach discussed the wartime work of the

Emergency Committee in Psychology, which had operated under the auspices of the National Academy of Sciences. Charles Bray did the same for OSRD's Applied Psychology Panel. Leonard W. Doob, another psychologist, presented the work of OWI's Overseas Branch. Yet another psychologist, Dorwin Cartwright, wrote about wartime social psychology projects. And the sociologist Paul Lazarsfeld wrote a lengthy expository review of the *American Soldier* study.[22] Many other authors also offered favorable comments on wartime research, including a piece about the practical value of the social sciences written by Frederick Osborn, a well-known eugenicist who had served as a general during the war, and another study about the use of the social sciences in the federal government from the Russell Sage Foundation.[23] A couple of synthetic studies provided a broader perspective on wartime social science. In his ill-fated essay for the SSRC, Talcott Parsons included a long section titled "Some Practical Achievements: The War Record" as part of his case for including the social sciences in the proposed NSF. *The Proper Study of Mankind*, written by the prolific journalist Stuart Chase, drew special attention to social scientists' wartime contributions as well.[24]

Chase's study, like Parsons's essay, has special significance because it too was commissioned with the specific aim of improving the public image of the social sciences during the difficult postwar years. In this case, representatives from the SSRC and the Carnegie Corporation asked Chase to write a book informing the reader about "a very respectable body of social science," which "works," which "predicts events to a degree," and which "came of age during the war." After agreeing to undertake this project, Chase emphasized that his analysis would look "forward not backward." That is, he would present wartime social research as a model for solving practical problems in the postwar era.[25] For his contribution, the SSRC gave Chase $7,500, a sizable payment but less than the $10,000 Parsons received.[26] Moreover, after realizing that Parsons's manuscript would need "rather drastic revision," SSRC president Pendleton Herring said he hoped Chase's book would be "widely read by the public."[27] Indeed, unlike Parsons's stillborn piece, Chase's work became a national best seller and gained even greater exposure through printings in other countries.[28]

In Chase's account and in the other social science war stories, certain themes appear frequently, starting with the declaration that World War Two gave a decisive boost to the scientific character of social research. Consider Lazarsfeld's review of the *American Soldier* study, which conveys a thoroughly positive message, unlike the analyses presented by Lynd, Glazer, and Schlesinger. In a glowing appraisal, Lazarsfeld, a leading scholar in the statistical study of attitudes and public opinion research, found many things to admire about the study, starting with its immense scale. More than 600,000 interviews, conducted by the Research Branch of the Army's Information and Education Division, comprised the project's raw data, and its results appeared in four substantial volumes. The study thus had no "parallel in the history of the social sciences," proclaimed Lazarsfeld. "Never before" had "so many aspects of human life been studied so systematically and comprehensively."[29] In an account of military psychology, a field that had barely existed in 1940, Charles Bray similarly heralded the impressive scientific character of wartime work. Bray did admit that psychologists focused on very practical matters during the war and thus "contributed little to an increased understanding of the human being." Nevertheless, he then underlined the positive point that their work had relied on "experimental evidence" and was "based on measurement."[30]

Also, contrary to the critics' accounts, these favorable assessments commended wartime efforts for promoting a type of social science that produced results of much greater value compared with the more speculative and less scientific forms of social inquiry common in the past. Perhaps nobody made this point better than Stuart Chase. Long enchanted by the hope that social science would pave the way for a better world, Chase had been enamored with Soviet-style economic planning during the 1930s. But in the postwar years he rejected that vision. Despite the scientific aspirations of Karl Marx and his scholarly followers, Chase now associated their work with an older, outdated, and philosophical or "arm-chair" approach. Instead, he favored studies that posed questions whose answers depended on rigorous empirical social inquiry, which, Chase added, had risen to prominence only recently, especially during World War Two. As a result, hard-nosed researchers could now "answer certain problems of society better than any dictator, better than a convocation of elders, better than intuition or common sense."[31]

Practically oriented, empirical, and quantitative social research carried out during the war also gained support in these accounts because such work seemed to fulfill military needs effectively. In recounting his work as director of the *American Soldier* study, former general Frederick Osborn explained that social scientists had succeeded in getting the "immediate and interested attention" of their wartime superiors only when they presented "a carefully authenticated array of facts, clarified by simple charts, indicating what might be expected to happen under various alternatives." This approach helped secure the army's acceptance of the well-known point system of discharge. Under this system, soldiers received points based on certain criteria, such as how long they had served. Their totals then determined the order in which they returned home at the war's end.[32] Praise for social science work on this point system also came from Talcott Parsons, who went so far as to propose that its adoption "perhaps saved the country from . . . a serious morale crisis," which could have ensued if "overseas veterans had been kept on service while more recently drafted troops at home were discharged."[33]

By depicting wartime social research in these ways, scholars and their allies including Chase and Osborn also implied a fundamental likeness between the social and natural sciences. Had wartime physics produced new theories of cosmic evolution or fundamental forces? No. Instead, physicists had concentrated on technical even if sometimes monumental challenges, such as building atomic weapons or developing radar systems for detecting enemy submarines—though prior advances in theoretical physics provided the basis for such work. Similarly, wartime social science consisted primarily "not of new research but of the application of already existing knowledge to practical problems," claimed the Russell Sage Foundation in its 1950 report on effective uses of the social sciences by federal agencies. The war had thus encouraged social scientists to become "social practitioners," similar to "engineers and other specialized technologists" who applied "knowledge of the physical sciences."[34] Lazarsfeld similarly applauded the *American Soldier* study for demonstrating how "social engineering" could overcome "objectively difficult" situations.[35]

And so it went, with social scientists and their supporters making the same set of points time and time again: they deserved recognition and support based not on some vague promise of future progress but on their recently demonstrated success in developing practically useful and scientifically reliable results during wartime. These claims fit a broader pattern, which has been noted by Ron Robin and other historians, wherein social scientists mobilized by the war argued in the postwar years that they would "provide objective methods for controlling human behavior."[36] Just as importantly, their wartime accounts downplayed any concerns about the political control of social research and its ethically questionable uses. No radicals, critical theorists, or humanistically inclined scholars graced the pages of the accounts by Chase, Parsons, Bray, Leighton, and others—only hard-nosed, scientifically minded scholars working on specific problems defined by their wartime superiors.

In an excellent study of think tanks in the twentieth century, James A. Smith has claimed that "social scientists thus emerged from the war with greater confidence and their reputations enhanced."[37] No doubt, many of those scholars sincerely believed in the importance of their wartime work, and they saw great postwar opportunities to advance their profession by building on their wartime experiences. However, in light of the difficulties facing social scientists and federal agencies examined earlier in this section and in the previous chapter, social science war stories should not be taken at face value. Moreover, my analysis has revealed that besides constituting a frustrating episode for the social sciences, the NSF debate showcased a more widespread scientistic strategy for gaining recognition, acceptance, and funding for the social sciences during the early Cold War years. Meanwhile, the claim that any wartime or postwar social science research efforts had a normative orientation and, even more specifically, a leftish tilt remained for conservative critics to present. Only many decades later would Barry Katz and other scholars begin to analyze the experiences of left-leaning scholars in wartime agencies, including members of the Frankfurt school and some homegrown radical economists in the OSS.

Undernourished and Underutilized: RDB's Human Resources Committee

While their war stories suggested that military-funded research should provide a model for future progress, social scientists often encountered difficulties in their efforts to secure military funding during the early Cold War years. In 1946, the sociologist William Ogburn recommended that "for every subsidized piece of research in natural science there should be corresponding financial aids to research in social science."[38] But such a proposal had little chance to succeed. In fact, three years later, when the military had already become a towering presence in the Cold War science patronage system, the social sciences and psychology together received only 1.5 percent of the military R&D budget.[39] In order to illustrate what this marginality meant to the social sciences, this section discusses the development of the Human Resources Committee within the military's Research and Development Board (RDB).

Created by the same 1947 legislation that unified the armed forces in the National Military Establishment (soon renamed the Department of Defense), the RDB reported to the U.S. secretary of defense and was responsible for developing an integrated scientific program to serve national security needs. Vannevar Bush became the board's first chairman. In the next few years, Bush's successors included the MIT president and physicist Karl Compton, the MIT trustee and chemist William Webster, and then the MIT chemical engineer Walter G. Whitman, while another physical scientist, Lloyd V. Berkner, served as RDB's executive secretary. By 1949, the board had a staff of 250. Through an extensive system of a hundred or so specialized committees, another 2,500 civilian and military personnel contributed to the board's work.[40]

As already suggested by the composition of RDB's leadership, social scientists were destined for a rough time here. Remember that Bush purposely left them aside in his 1945 landmark science policy report to the president, while his statements during the NSF debate recommended a cautious approach regarding their position in the proposed science agency. In addition, Compton played a leading role among scientists who opposed inclusion of the social sciences. When Berkner commented on the social sciences, he asked if they could be "made exact" and suggested that they should search for mathematically precise statements describing the relations connecting "the behavior of the brain and the consequent human response."[41] Bush, Compton, Webster, and Berkner also had close connections to one another. All of them were members of the prestigious National Academy of Sciences (NAS), which included the social sciences only in a limited way, and which placed special emphasis on fields that overlapped with the natural sciences and seemed free from worrisome political associations. Webster and Compton worked in the wartime OSRD under Bush as well, while Bush chose Berkner as his first assistant at the OSRD. Of the social sciences, only psychology enjoyed a significant presence in the OSRD, through its Applied Psychology Panel, as noted in the previous chapter. Bush, Compton, Webster, and Whitman also associated with one another through their MIT positions. And before midcentury, MIT's social science activities remained rather limited.[42]

Yet social scientists and psychologists did have some presence through RDB's Human Resources Committee (HRC). Though it had no power to allocate money, the HRC provided budgetary recommendations regarding RDB's social science efforts, and it evaluated specific research projects. Like other RDB committees, the HRC included civilian and military personnel and had a civilian leader, in this case Donald Marquis. One of the most prominent psychologists in postwar America, Marquis became the chairman of the University of Michigan's Psychology Department, served as the president of the American Psychological Association (1947), and, as will be discussed in chapter 3, had important roles in the development of the Ford Foundation's Behavioral Sciences Program. Four specialized panels helped the HRC carry out its mission, each of them also led by a highly accomplished chairman who had extensive experience with military projects. The Psychophysiology Panel was directed by the psychologist Lyle H. Lanier; the Personnel and Training Panel by another psychologist, Robert L. Thorndike; the Manpower Panel by the demographer Philip Hauser; and the Human Relations and Morale Panel by the sociologist Charles Dollard.[43]

As one would expect, the HRC, a social science committee, got off to a dicey start. At HRC's first meeting, Vannevar Bush described its field of study as "unusual." "Quite frankly," said Bush, he viewed this particular committee as "an experiment." While most other RDB committees would address "clear-cut" problems," HRC's field of interest seemed "very amorphous by comparison," for work in the social sciences had an "intangible nature." On other occasions Bush conveyed caution toward these sciences but not always explicit criticism. In this setting, however, he added that the social science community included a "lunatic fringe."[44]

Less than a year after that discouraging start, an HRC statement observed that concerns about its social science activities had arisen elsewhere within the federal government as well. Specifically, the Republican senator from New Hampshire, H. Styles Bridges, had asked the Senate Appropriations Committee about the value of an HRC study called "Investigations in the Differences in Readership and Listenership Habits." Though Bridges apparently challenged only this one research project, his skepticism seems to have been an extension of wider conservative attacks directed at the OWI and later at the Voice of America. The HRC statement also noted that the Office of Naval Research had asked about the propriety of funding HRC's social research projects more generally.[45]

Soon thereafter the HRC highlighted a series of problems plaguing the military's social science effort. In its 1948 report "Research and Development in Human Resources in the National Military Establishment," the HRC complained that its funding that year amounted to little more than one percent of RDB's entire budget. On top of this limitation, HRC noted that its relatively small budget supported "developmental and 'service' activities" that contributed "only slightly toward general improvement in methods of utilizing human resources." More general problems in the military's social science efforts included "the small number of professional personnel engaged, the lack of adequate research facilities and the virtual absence of coordinated programs directed toward long-range objectives." At the very least, additional funding seemed warranted, enough to increase HRC's share of RDB's overall budget to 5 percent, or about $20 million, plus an additional $2 million for a social science graduate training program, similar to a program recently developed for the natural sciences by the Atomic Energy Commission.[46]

One year later the HRC complained again about severely inadequate funding. Its estimated 1950 budget remained under $10 million, not even 2 percent of the military's total R&D budget. A recent increase in air force support for social research represented a step forward, though even here the additional funds represented only three-fourths of what HRC had requested. Meanwhile, the other two military branches had "reduced or barely maintained their level of support," with the navy providing about 25 percent of HRC's recommended level and the army 30 percent. Also worrisome, HRC's proposed graduate training program had not gone forward. The overall result seemed deplorable: the HRC declared that in order to succeed, its efforts to confront "an expanding range of problems under conditions of great urgency" required much greater support. Under existing conditions, with no more than "a skeletal force," the committee said it simply could not fulfill its national security responsibilities effectively.[47]

Perhaps most galling of all, the HRC concluded that lack of adequate support for the social sciences did not reflect a more general lack of funding for the national defense effort and its scientific activities. With the U.S. defense budget having recently reached $15 billion, one HRC report indicated that its members were "unwilling to believe" that the defense establishment could not find $15 million for research on the "human factor." "That is, the Committee is still not willing to accept the fact that one-tenth of one percent of the defense budget should not be spent for research" in this area.[48] During the Korean War, the defense science budget would soar even higher, tripling between 1950 and 1952, from more than $500 million to $1.6 billion.[49] Under these conditions, social scientists could hope that the growing horn of plenty would benefit everyone.

As it turned out, however, the HRC found little reason to rejoice. Owing to insufficient incentives, "the loss of potential scientists to the human sciences and related fields has approached the danger point," warned a 1951 memorandum to the RDB chairman. With the fields of advanced social science training "being depleted," nothing less than "the future of the nation" seemed to be "at stake."[50] Unfortunately, an HRC working group had found "only limited understanding" within the military services of "the benefits that can accrue through the application of the results of social sciences research to the military arts." Thus "old established military customs, procedures, and policies" remained in place, even though they had no basis in "scientific fact or theory."[51]

As this discussion of HRC has revealed, social scientists struggled for dollars, respect, and influence within the military science system in the early Cold War years. Following the pattern established during World War Two, these scholars quickly found themselves on the sidelines during the late 1940s and early 1950s. Based on their experiences with the RDB, HRC's prominent social scientists argued that their fields suffered gravely from undernourishment and insufficient use. In their estimation, such glaring deficiencies demanded urgent national attention, a point made as forcefully as possible by the HRC working group noted above: "OUR NATIONAL HUMAN RESOURCES CAN ONLY BE EXPLOITED WITH MAXIMUM EFFICIENCY BY CLOSE TEAMWORK BETWEEN THE EXECUTIVE MILITARY BRANCHES OF THE ARMED SERVICES AND ALL BRANCHES OF SOCIAL SCIENCE RESEARCH."[52] Adding insult to injury, some legislators continued to raise doubts about the value of social sciences and their support from the military. In 1953, the HRC executive director reported that military programs had, unfortunately, come under congressional scrutiny because "several" legislators had "no realistic conception of what research in this field really involves."[53]

Under these conditions, military funding for the social sciences would never come close to the level of support for the physical sciences. No doubt one cause of this disparity was the fact that defense-oriented physical science research often required extensive equipment and large research teams and thus had become terribly expensive. But the disparity in funding also reflected pivotal developments in American partisan politics and national science policy that made questions about the nature and value of the social sciences troublesome at every turn, that provoked political scrutiny of federal agencies involved with these sciences, and that ensured that the social sciences remained a peripheral matter of concern within the defense science establishment as well as within the federal science system more widely. During the 1930s, when social scientists had significant roles in New Deal agencies and when natural scientists did not yet dominate the federal science system, the distribution of funding had been significantly different. According to one government report from the mid-1960s, in 1938, nearly 25 percent of all federal science dollars went to the social sciences.[54] But a contemporary assessment in the early 1950s noted that the physical sciences received more than 70 percent of all federal science funds, and the life sciences received nearly 20 percent. Meanwhile, the social sciences' meager share stood at 3 percent, while other fields received the remaining 7 percent.[55]

New Patrons and Research Organizations

Despite the social sciences' peripheral status, the military still became an increasingly important patron for them during the Cold War years. By midcentury military funding already amounted to millions of dollars each year. Furthermore, with the founding of new military research organizations, funding became more robust. At the same time, scholars and their advocates continued to emphasize that the military–social science partnership should concentrate on advancing research defined in scientistic terms.

In 1946, the navy created the Office of Naval Research (ONR), which acquired an especially good reputation among the nation's scientists. Besides providing substantial research funding, the ONR made an effort to accommodate the needs of academic scientists. The agency thus supported a large number of basic science projects, especially during its early years. According to one midcentury report, the navy's premiere science organization became "the federal government's first peacetime effort to support basic research in all fields of science on a large-scale basis."[56] As the postwar NSF debate dragged on into the late 1940s, many scientists welcomed the ONR as a surrogate for the proposed civilian science agency. Though the ONR focused predominantly on the natural sciences, it established a Human Resources Division to support the social and psychological sciences.[57]

The army created new social science organizations to serve its needs as well. These included the Human Resources Research Office, founded in 1949 under a contract with George Washington University.[58] During the Korean War, these organizations studied such matters as the training of soldiers, motivation, group dynamics of combat units, and logistical challenges, all subjects that had received significant military support during World War Two.[59] A few years later, in 1956, the army established the Special Operations Research Office (SORO) under a contract with American University. SORONs, as the organization's researchers playfully called themselves, produced handbooks on foreign areas with information about social structures, political systems, and estimated revolutionary potential. The SORO also developed a Counterinsurgency Information Analysis Center and soon became a major military center for counterinsurgency research.[60]

Meanwhile, the air force set up a number of units that supported social and psychological research. Its Human Resources Research Center and Human Resources Research Laboratory focused narrowly on problems of military personnel and thus carried out studies on such issues as personnel training and personnel classification. But the air force's Human Resources Research Institute (HRRI), founded in 1949, had broader responsibilities and provided support for the massive Refugee Interview Project, which was carried out by Harvard's Russian Research Center and will be discussed later in this chapter.[61] The air force established the most famous of the early Cold War think tanks as well, namely, the RAND Corporation— (from *R*esearch *AN*d *D*evelopment). Preceded by an air force–supported research unit located at first within the Douglas Aircraft Corporation, RAND became a legally independent, private, nonprofit scientific research organization in 1948. Though it acquired some valuable seed money from the Ford Foundation, RAND relied heavily on air force funding through the early 1960s.

RAND's involvement with the social sciences commenced with an initiative from John Williams, the first leader of its Mathematics Division, a noteworthy point given the think tank's subsequent role in fostering "the pervasive quantification of the social sciences," as the historian David Hounshell has written.[62] Williams, the author of a popular book on game theory, later joked that after some debate within RAND about the possibility of including the controversial social sciences, skeptics had sent him to Washington, D.C., to discuss the idea. Aware of the widespread coolness toward the social sciences within the Defense Department, his colleagues had assumed that the military brass would "kill the idea," at which point Williams would have had to "stop pestering people." To their surprise, however, General Curtis LeMay, the chief of the Air Force Staff for Research and Development, approved of his idea.[63]

Subsequently, RAND held a conference in New York to recruit social scientists. Warren Weaver, another mathematician who consulted for RAND, gave the plenary address. Both Williams and Weaver especially admired the wartime contributions of mathematically inclined economists. The New York conference included twelve panels grouped under four broad fields: psychology and sociology, political science, economics, intelligence and military affairs. A fifth area focused on methods, organization, and planning of research. Toward the end of the conference, the sociologist Herbert Goldhamer noted that RAND had been approaching the social sciences hesitantly, "tentatively circling around a rather vast ocean and getting its toes wet." But the conference had heightened interest, sending RAND into "the midst of the waves."[64]

Indeed, soon after the New York event, RAND hired the psychological warfare expert Hans Speier to lead its new Social Science Division and Charles J. Hitch, a former operations researcher at the OSS, to lead its new Economics Division.

During the early Cold War years, additional funding came from intelligence and propaganda agencies, which deserve brief note here because, though not part of the military science establishment, they helped define the wider context for federally funded research related to national security issues. Among the most important of these agencies was the Central Intelligence Agency (CIA). Created by the 1947 National Security Act, the CIA became a massive operation that focused on fighting the Soviet enemy and its influence around the world through intelligence analysis, espionage, and other secret political actions. Like the OSS, its predecessor, the CIA depended heavily on university scholars, including many from the social sciences. The CIA also had an Office of Scientific Intelligence. Dominated by physical scientists, this office made dramatic advances in spy satellites and the analysis of enemy military capabilities. It even carried out some controversial research on psychic powers, hypnosis, behavioral modification, brainwashing, drugs including LSD, and other forms of mind control.[65] However, in the CIA's other original directorates for geography, history, and politics, social scientists did not have to answer to physical scientists, in contrast to the situation in the defense science establishment. Though at first economists did not have their own CIA directorate, they soon became prominent in the agency's intelligence analysis activities doing research on the Soviet economy and its military potential, for example.[66]

Further resources flowed through governmental agencies responsible for Cold War–oriented propaganda and information campaigns, with the difference between information and propaganda often murky. In his influential writings that informed the development of containment theory, George Kennan argued that the Kremlin sought to fill "every nook and cranny available to it in the basin of world power." America thus needed to take the ideological offensive, in order to create "among the peoples of the world . . . the impression of a country which knows what it wants, which is coping successfully with the problems of its internal life and with the responsibilities of a World Power, and which has a spiritual vitality capable of holding its own among the major ideological currents of the time."[67] This line of reasoning informed the 1948 Information and Educational Exchange Act, which supported U.S. information programs overseas; the short-lived Psychological Strategy Board (1951–1953), created in the executive branch to plan and coordinate psychological warfare activities; and the U.S. Information Agency (f. 1953), which promoted a sympathetic understanding of American interests in other countries around the world. Agencies involved in propaganda, psychological warfare, and information activities also found it necessary to carry out research to support their missions.[68]

Funding provided by all the agencies noted above had major implications for the growth and trajectory of the social sciences in Cold War America. "Military, intelligence, and propaganda agencies provided by far the largest part of the funds for large research projects in the social sciences in the United States from World War II until well into the 1960s," as the historian of communication studies Christopher Simpson has pointed out.[69] Equally important, scholars who worked with national security–oriented agencies and especially with defense science agencies regularly claimed that social and psychological research had certain readily identifiable characteristics that placed it on the forefront of scientific progress, much like those scholars and their supporters who produced the favorable accounts of wartime social science considered earlier. John Darley, a psychologist who served as the chairman of ONR's Advisory Panel on Human Relations and Morale, presented a typical discussion. After a conference about the first five years of ONR-sponsored research on group behavior, leadership, and individual behavior, Darley presented an admiring report, claiming that in this research one found "not a transcendent or pretentious" social science, but the "slow, arduous, painstaking processes of science generally." He explained that successful ONR research proposals concentrated on "observation and delimitation, preliminary investigation, formulation and testing of hypotheses, correction and reformulation." The resulting research put forth "tentative and limited" generalizations, not recklessly bold claims, and therefore exhibited "a notable absence of intuitive and sweeping conclusions." These ONR-sponsored studies thus "followed general standards of evidence and proof via the methods of statistical test, replicated experiments, and predictive outcome." On the flip side, Darley declared that the ONR had no interest in many approaches now considered to be insufficiently scientific, including "traditional economics, political science, and history."[70]

Social Science Weapons of War

Yet one might still wonder if Darley's claims about ONR-sponsored social research served mainly as window dressing. After all, beginning in the early postwar years, SSRC's social scientists and their supporters including the journalist Stuart Chase regularly made such claims at least partly to advance a vision of social science that could appeal to skeptics in the scientific, military, and political spheres. Perusal of the edited volume of ONR-supported studies published after the 1951 conference mentioned above indicates that a quantitative, experimental, and predictive orientation did, indeed, predominate. But to appreciate the pervasiveness of the scientistic outlook in military-sponsored research programs, and to understand how that outlook became manifest in particular fields of inquiry, requires a different level of analysis. This section considers the development of three fields of inquiry that received extensive military support, that became influential in the scholarly arena, and that also acquired practical influence on the Cold War battlefield: communications studies, the decision sciences, and the science of strategy.

Communications Studies

The first field of inquiry, communications studies, had roots in the interwar era and originally included a line of critical scholarship promoted mainly by liberal scholars, including prominent social scientists who wanted to direct scientific social inquiry toward progressive ends. These scholars worried that the dissemination of propaganda through powerful new forms of mass media threatened democracy. However, the emergence of an unfriendly political climate, conservative criticism, and a lack of adequate funding helped to undermine that effort. As the project of examining the dangers of propaganda went out of fashion, research on communication effects within a purportedly value-neutral and instrumentalist framework received much greater support from a combination of sources, including the military.[71]

Participants in the post–World War One debate about the social role of propaganda included many left-liberal intellectuals and social scientists, such as Charles Beard, John Dewey, Leonard Doob, Walter Lipmann, and Robert Lynd. As they saw it, the government's control of wartime information and communications, especially through the U.S. Committee on Public Information, had dire implications for American democracy. In the next two decades, these concerned voices warned that the tremendous power of modern mass communications associated with the rise of radio and film facilitated a dangerous manipulation of the masses by political leaders from Roosevelt to Hitler and by corporate commercial enterprises.[72]

In the mid-1930s liberal researchers interested in developing this critical perspective, often with the help of detailed case studies, gained support from the Institute for Propaganda Analysis (IPA). Located in New York City, IPA's advisory board included Beard, Doob, and Lynd. In his 1935 book *Propaganda: Its Psychology and Technique*, Doob explained the "social philosophy" embedded in this work: he and others hoped that "the recognition and understanding" of propaganda would enable the reader "to free himself to a certain extent from the forces which that phenomenon represents."[73]

Meanwhile, an opposing research approach focused on how national leaders and other elites could control the instruments of mass media for constructive purposes. Leaders of this approach took propaganda as a fact of mass societies, and not necessarily a regrettable fact. They then proposed that governments and other reputable institutions of modern society, with the help of communications experts, could use propaganda to influence public opinion and associated behaviors in desirable ways. According to Wilbur Schramm, a remarkably influential figure in this emerging field, the effective use of mass communications could strengthen social stability in modern societies. He explained that modern society, unlike the typically "small" and "relatively homogenous" community of the not-too-distant past, was "heterogeneous and massive." Its residents thus had to rely on groups to represent their interests. Herein lay a major problem, for these groups tended to be selfish and thus advanced conflicting interests, which produced harmful social conflict rather than desirable social integration. However, through the skillful use of "the large media," leaders in modern society could cut through these competing differences, Schramm argued. In doing so, they would facilitate "the exchange of opinion and the creation of social consensus."[74]

According to this vision, communications experts would play a major role in ensuring social progress. Through their scientific research they would reveal how to use various types of mass media to influence target populations in the anticipated fashion. As the political scientist Harold Lasswell famously put it, their scholarship would indicate "who says what in which channel to whom with what effect."[75] To its advocates, this instrumental conception of communications studies pointed the way toward a healthy integration between purportedly apolitical scientific progress and the practical applications of the resulting knowledge.

During World War Two and the early Cold War years, partisan politics, institutional developments, and funding patterns decisively shifted the balance of influence in favor of the instrumentalist perspective. In yet another attack on liberal social science and its patrons, congressional conservatives challenged the IPA and helped bring about its demise. When the institute closed in 1942, the critical analysis of propaganda lost its major institutional basis of support. Meanwhile, the alternative approach, now deployed to shape beliefs, attitudes, and actions on a wide range of target populations both at home and abroad, attracted substantial funding, in the 1930s from the Rockefeller Foundation and Carnegie Corporation, and in the 1940s and 1950s from additional sources including the military; other national security–oriented agencies such as the CIA, the Voice of America, and the Psychological Strategy Board; and the modern Ford Foundation.[76]

As explained by Edward W. Barrett, President Truman's assistant secretary of state for public affairs from 1950 to 1952, this field of study had great value on the Cold War battlefield. According to his rather dramatic account, "totalitarian tyrants" had surged "miles ahead" of the United States in recognizing the "growing force of mass opinion." Unless Americans intended to commit "suicide," the nation's leaders would need "to master the techniques of international persuasion" through the skillful use of mass communications. Fortunately, the nation had made a good start, Barrett added. To get the American viewpoint across, government operations already employed a wide range of media, including "leaflets, posters, magazines, films, lectures, books, exhibits, libraries, information centers, exchange of students and editors and teachers and leaders, press releases, day-to-day contact with commentators and other opinion-formers, scientific newsletters—in brief, every medium that can sway men's minds."[77]

For anyone seeking to better understand the combination of scientific aspirations and strategic promise that suffused a great deal of work in communications studies during these years, Project Revere provides a good starting point. This project, which proceeded under contract at the University of Washington, received extensive support, totaling more than $300,000, from the air force's HRRI and the CIA-funded Society for the Investigation for Human Ecology. The project's leader, Stuart Dodd, was a mathematically oriented sociologist who had been the U.S. Army's director of surveys in Sicily during World War Two. According to an influential 1958 book by two other sociologists involved in this study, Melvin DeFleur and Otto Larsen, Project Revere had two main scholarly objectives: first, to study "the effects of the flow of information" on variables such as "repetition and redundancy," and second, to test "the feasibility of tracing the flow of information through communication networks which operated to diffuse the messages beyond those persons who were directly exposed." In pursuit of these goals, Revere's research team used airplanes to drop "approximately 750,000 leaflets" containing more than fifty message versions on "35 unwarned communities." Under military orders, media outlets in these communities refrained from commenting on the deluge of leaflets while the experiments were underway. In the biggest of these experiments, airplanes dropped 326,000 leaflets on Birmingham, Alabama, or about one leaflet for each unsuspecting city resident.[78]

In their discussion, DeFleur and Larsen emphasized the leaflet's special scientific value. Simple in design and easy to modify, the leaflet permitted social scientists to gain "a great deal of experimental control." The sheer volume of leaflets that could be dropped in combination with the careful manipulation of relevant variables, such as message content, facilitated the use of quantitative techniques of analysis as well. According to the original research plans, researchers would check the scientific findings of leaflet studies carried out on American soil against the results of studies carried out in other countries, for example, in "Japan, Korea, North Africa, Spain, etc."—another indication of the lofty scientific ambitions motivating this study.[79]

FIGURE 4. Preparing a Korean leaflet bomb. At the Far Eastern Command Printing Plant, Yokohama, Japan, SFC Furl A. Krebs loads an M16M1 cluster adapter with 22,500 5 × 8 inch psychological warfare leaflets. November 1, 1950. *U.S. Army photograph. Naval History and Heritage Command.*

Beyond the anticipated scientific results, the practical motivation of Project Revere was to determine how the American military could stimulate "the target population to take some appropriate action," as the allusion to the American revolutionary hero Paul Revere suggests. More specifically, the air force sponsor wanted to learn how it could use the leaflet as a "weapon," to "set in motion social processes of persuasion that will direct and affect the behavior of mass populations" in American-friendly directions. The study's leaflets included a variety of messages requesting that their readers take particular actions, including one that specified how to donate blood. But "in an actual Air Force operation," DeFleur and Larsen continued, "compliance behavior" would include following "directions for surrender, sabotage, safety, or some other act."[80]

Though this project stands out because of its large size and vaulting ambition, Revere's investigative style nicely represents the dominant scientistic and instrumentalist characteristics of communications studies during the early Cold War era. The field's researchers often focused on perfecting, as DeFleur and Larsen put it, "systems of order and control," while the supporting research rested on statistical and experimental foundations.[81] Whereas scholars associated with the IPA had worried that "propaganda is evil and tricky," the typical researcher in this field now focused on "how propaganda functions," as Leonard Doob observed. These social scientists aimed to be neutral, not critical. Their work combined anticipated strategic uses with scientific knowledge production in a seemingly airtight package.[82] And as Christopher Simpson has argued in his smartly titled book *Science of Coercion*, for

"U.S. military, propaganda, and intelligence agencies," this rapidly growing research field facilitated the use of "mass communication as an instrument for persuading or dominating targeted groups."[83]

A telling shift in language accompanied the evolution of communications studies during those years as well. After the term "propaganda" had acquired a negative connotation, scholars adopted more "polite terms," including "communication" or "information," because, as Doob explained, such terms seemed to "imply no value judgment." Having worked at the IPA and then the OWI, Doob was intimately familiar with the political, ideological, institutional, and financial components of this story.[84] This field of inquiry therefore became known as mass communications studies or communication research, rather than propaganda studies. Meanwhile, alternative approaches including critical propaganda research and qualitative case studies concerned with the organization, structure, and control of mass media became marginal. Furthermore, the field's mainstream now considered such approaches to be more reformist and value laden rather than strictly scientific.[85]

The Decision Sciences

Whereas the case discussed above shows how selective development influenced by the changing structure of patronage contributed to the prominence of scientism and a corresponding marginalization of alternative approaches in communications studies, the story of the decision sciences reveals that military funding also helped to stimulate important fields of study that scarcely existed before. Based on specialized mathematical techniques of analysis, the fields of operations research, systems analysis, and game theory all promised to improve the analytic process of decision making. The evolution of these fields also brought certain select areas of social science, especially economics, into close contact with the physical and mathematical sciences and thereby enhanced their standing within the defense science establishment. In the long run operations research, systems analysis, and game theory acquired importance in scholarly fields and practical activities far beyond the national security arena as well. Some evolutionary biologists took up game theory, for example, while systems analysis influenced a wide array of fields from family therapy to urban planning and business management.[86]

Operations research (OR) first flourished during World War Two, helping to improve the efficiency of Allied military operations through the analysis of technical problems involving the use of weapons and tactics.[87] For example, OR studies of aircraft antisubmarine attacks on German U-boats led the Allies to change the settings on depth charges. And research on German attacks on naval convoys prompted the Allies to increase the size of those convoys, which reduced average ship loss. As U.S. Navy admiral E. J. King later explained, the

Allies benefited immensely from operations researchers, owing to their "exacting analysis" of "the measures and countermeasures introduced at every stage by ourselves and the enemy."[88]

With major funding from the military science agencies and additional support from civilian science agencies and private foundations, OR expanded dramatically in the postwar years. The army established its Operations Research Office under contract at Johns Hopkins University. The navy had its own Operations Evaluation Group, under a similar arrangement with MIT. The air force had an in-house Operations Analysis Division and used RAND for contract research. The Joint Chiefs of Staff of the National Military Establishment relied on its own Weapons Systems Evaluation Group. In 1949 the NAS also created a Committee on Operations Research. By the early 1950s, some two-dozen business organizations carried out operations research under military contract as well. Deeply shaped by its wartime roots, OR continued during the Cold War to focus on quantitative measures for evaluating the effectiveness of military operations in such areas as antisubmarine warfare, guided missiles, radar, and atomic weapons.[89]

Thanks largely to the boost from national security patrons and academic collaborators, OR soon became a "firmly established profession." Major research universities including MIT, Columbia, and Johns Hopkins developed formal advanced training programs. Rapid growth in the number of trained personnel made it possible to create the first professional society in the United States. Founded in 1952, the Operations Research Society of America also established the field's first American journal. The society's first president was the MIT physicist and former director of the military's Weapons Systems Evaluation Group, Philip M. Morse. Together with George Kimball, a chemist and former deputy director of the navy's Operations Evaluations Group, Morse published the field's first major textbook, *Methods of Operations Research* (1951). As in many other cases, Morse and Kimball's textbook first appeared as a classified study and later was published in the open literature, in declassified form and "after suitable modification."[90]

During OR's early development physical scientists and mathematicians represented the dominant disciplines, but economists soon became important participants. As explained by the noted Harvard biochemist and early proponent of social systems analysis Lawrence Henderson, the growing influence of economists reflected the realization that complex questions about the efficient use of military weapons and other resources could not be evaluated in terms of "maximum performance from the technological point of view." Military managers needed to evaluate "how to get the maximum, in terms of some measure of military effectiveness . . . for a given expenditure or resources."[91]

In sum, OR offered a sterling example of how scholars from one social science discipline, economics, could advance their scientific credentials and acquire valuable resources by working with physical scientists and mathematicians. Not incidentally, of all the social science disciplines, economics embraced formal mathematic analysis and statistical inquiry most fully, a point underscored by Morse the physicist.[92] Furthermore, as historical accounts by Stephen Waring and others have shown and as the brief discussion above confirms, the "customer-consultant relationship between the national security state and the scientific discipline [OR] shaped the purposes and assumptions of the field," including its dominant concern with improving the technical effectiveness of specific operations given existing national security structures and specified military goals. Under these conditions, the quest to maintain what one introduction to the field called "the true scientific attitude" meant that operations researchers had to bracket or simply ignore the complex ethical and political questions concerning those very same military operations, structures, and goals.[93]

OR acquired additional importance because its scholars, techniques, and institutional supports contributed to other cutting-edge fields of study concerned with national security matters, with systems analysis having perhaps the closest relations. Whereas operations researchers typically began with a clear-cut goal and with the available weapons or other resources already specified by their military superiors, systems analysts often started with a goal defined within a more general framework of strategic concern and then had to determine the best type of weapons or even the best weapon system for reaching that goal. Once again, the emphasis on quantitative cost-benefit analysis of alternative possibilities helped make economists leaders in this new field, while RAND became the field's central home during the 1950s.[94]

The classic example of systems theory comes from an early 1950s study of overseas air bases carried out by a group of scholars led by Albert Wohlstetter, a mathematician in RAND's Economics Division. Wohlstetter had studied symbolic logic at Columbia University, worked for the government on the application of mathematical tools to economic analysis in the 1930s, and then became involved in government studies of quality control in the production of military supplies during World War Two. Shortly after arriving at RAND, he joined an OR group studying appropriate sites for overseas air bases. But, going beyond the goal of determining the most appropriate overseas sites, Wohlstetter began to think about the larger Cold War strategic framework and various types of potential Soviet attacks. Though common wisdom supported the nation's heavy reliance on overseas air bases as a sound strategy for deterring Soviet aggression, Wohlstetter now disagreed. He argued that it would be more efficient from an operational viewpoint, more effective in light of strategic aims, and also less costly to keep American planes on bases inside the United States, while using overseas bases for repairs and refueling. Consequently, his analysis did not lead him to specify the best means for achieving the original objective. Instead, he reconceived the problem in light of wider considerations and a more complicated array of factors. What began as an OR problem turned into a type of systems analysis.[95]

The development of systems analysis during the early Cold War years drew on a wide variety of fields, including cybernetics, economic cost-benefit analysis, game theory, electrical engineering, mathematics, and business management. A shared outlook emphasizing the importance of the system over and above its component parts united the nascent field's diverse researchers. "Systems Analysis deals with quantitative interrelationships between the important variables determining system performance," noted one scholar.[96] As a military report on the design and use of "Man-Machine Systems" explained, researchers used systems analysis to determine how "complexes of men, machines and [their] multiple interactions can be efficiently shaped, controlled, and their outcomes predicted." Ideally, this work described in "relatively few quantitative terms not only how an existing system behaves, but also how to improve or optimize, in terms of tradeoffs, its design, development, and change characteristics during its life cycle."[97]

Game theory offered a third cutting-edge line of social inquiry that aimed at improving decision making through rigorous logical analysis and quantitative reasoning. Toward the end of World War Two John Von Neumann and Oskar Morgenstern published the field's foundational study, *Theory of Games and Economic Behavior* (1944). This monumental work of more than six hundred pages offered "an exposition and various applications of a mathematical theory of games." Von Neumann, a physicist, a professor at Princeton's Institute for Advanced Study, and a prominent national security adviser, had worked on this project with Morgenstern, a Princeton professor of economics, since 1940. Their purpose certainly did not lack ambition, as they aspired to build the foundations for a "new approach to a number of economic questions as yet unsettled."[98] How did they propose to pull off such a feat?

"The nature of the problems investigated and the techniques employed," claimed the two Princeton scholars, required "a procedure which in many instances is thoroughly mathematical," including "mathematical logics, set theory and functional analysis." In a section outlining their "Methodological Stance," they declared they would be following the "best examples of theoretical physics." After considering the well-known objection that "economic theory cannot be modeled after physics since it is a science of social, of human phenomena," they labeled such a view "premature." Other scholars in this field made similar claims, with John Williams, the mathematician who helped bring social scientists into RAND, declaring with a dash of bravado that "Game Theory is very similar in spirit to the Theory of Gravitation."[99]

As the title of Von Neumann and Morgenstern's foundational text suggests, the analysis of a game lay at heart of this new field, particularly games like poker in which competing players develop strategies based on incomplete information. "The essence of a 'game,'" explained another prominent RAND game theorist named Martin Shubik, "is that it involves decision-makers with different goals or objectives whose fates are intertwined." From the analytic viewpoint, a game consisted of "the players or individual decision makers," "the payoffs or the values assigned to the outcomes of the game," "the rules which specify the variables that each player controls," "the information conditions," and "all other relevant aspects of the environment." Researchers then constructed models to determine the hypothetical outcomes of different strategies selected by the players, with the results determining the best strategy for playing any particular game.[100] This quantitative approach, added Morgenstern, aimed to "give precision to the notion of 'rational behavior,'" in contrast to "qualitative and philosophical arguments" that in his blunt evaluation had "led nowhere."[101]

To proponents of game theory who were inclined to think about strategic matters, the Cold War itself seemed to be one of these real-life games, in a wonderfully simplistic even if immensely disturbing form. This game consisted of two main players, the United States and the USSR, and conflict defined their relationship. Determining the best strategy, meaning the most instrumentally rational course of action, depended on how one expected the enemy to behave. One could thus use game theory's techniques to address fundamental strategic questions of the nuclear age, such as deciding whether it would be "rational" to launch a first strike on the enemy's military targets.

Though Von Neumann and Morgenstern originally highlighted the relevance of game theory to economics, its subsequent development depended much on its perceived military value. During the Cold War's first decade or so, defense agencies played a large role in the field's evolution. RAND and the ONR in particular, together with associated scholarly groups at major universities such as Michigan and Princeton, turned game theory into a vigorous intellectual pursuit, though this development did not happen overnight. Shubik estimated that when his book *Readings in Game Theory and Political Behavior* appeared in 1954, the literature included only about "ten or fifteen good articles on the subject relevant to the behavioral sciences." But ten years later, when his book *Game Theory and Related Approaches to Social Behavior* appeared, he could count "well over a thousand books and articles involving game theory; several hundred addressed to the behavioral sciences."[102]

The Science of Strategy

The final field considered here, which I'm calling the science of strategy, involved deterrence theory and counterinsurgency research. In the late 1950s and early 1960s, shifting perspectives on these topics contributed to a reorientation in national security strategy, from a stance that focused on nuclear deterrence based on the threat of massive retaliation to a position that placed increased emphasis on more conventional forms of warfare and counterinsurgency efforts.[103] Furthermore, the scholarly underpinnings of this policy shift moved beyond economic tools of analysis to include a wider range of psychological, sociological, and political studies. As Ron Robin has shown, during the early 1950s, proponents of these "behavioral science" studies at RAND had been "patronized by their peers in the physics and economics divisions for their lack of scientific rigor," leaving them to concentrate on strategies for fighting conventional warfare.[104] However, by the decade's end, as American concerns about communist-inspired insurgencies in developing nations reached a feverish pitch, behavioral science counterinsurgency studies acquired increasing influence within national security circles.

Clemenceau, the legendary nineteenth-century French military strategist, famously observed that war is "too important to be left to the generals." Writing in the early post–World War Two years, the University of Chicago–trained political scientist Bernard Brodie thought the generals still had far too much influence. After his wartime posts in the Bureau of Ordnance and the Office of Naval Intelligence, Brodie took a position in Yale's Department of International Relations. His 1946 book *The Absolute Weapon* and a stream of other publications made him a leading nuclear strategist and led to a new job at RAND's Social Science Department in the early 1950s. As Brodie saw it, the advent of nuclear weapons had fundamentally changed the nature of warfare and thus made the development of a new "science" of strategy essential.[105] But what exactly could it mean to subject the study of strategy to scientific treatment?

Though trained in political science, Brodie turned to economics for guidance, finding much "similarity" in the "objectives" of economic analysis and military strategy. If economic analysis focused on the "efficient allocation of the national (or other community) resources for the economic ends set down by the community," he reasoned that military strategy concerned "how the resources of the nation, material and human, can be developed and utilized for the end of maximizing the total effectiveness of the nation in war." From this perspective, the new science of strategy would embrace a "genuine analytic method" of the sort found in economics, a line of thinking increasingly common in military science circles and steadfastly opposed to what Brodie called the military man's "anti-theoretical" and "anti-intellectual" bias.[106]

In the late 1940s and 1950s, the effort to place military strategy on a more rigorous, analytic foundation flourished under the guidance of a new breed of civilian military strategists, just as Brodie anticipated. These strategists included researchers from the fields of operations research, systems analysis, and game theory. National security agencies including RAND, together with the large private foundations, provided bountiful resources for scholars and university programs, mainly political science departments and strategic studies institutes. Prominent programs flourished at Harvard and Princeton as well as the Institute for Defense Analysis, a consortium of five universities, Columbia, MIT, Pennsylvania, Georgetown, and Johns Hopkins.[107]

Pondering the threat of a surprise Soviet nuclear attack, civilian military strategists underscored the need to develop and maintain a powerful retaliatory force as a deterrent. In the early 1950s Brodie argued that the hydrogen bomb should be developed for this purpose, a position that helped justify the Eisenhower administration's policy of massive retaliation. The logic seemed straightforward: knowing the United States had such a powerful weapon would deter Moscow from launching a nuclear first strike. But by the decade's end, the problems associated with establishing a credible deterrent had shifted, leading the RAND systems theorist Albert Wohlstetter to claim that the "thermonuclear balance" had become "precarious."[108]

In his classic 1959 article called "The Delicate Balance of Terror," Wohlstetter argued that maintaining an effective deterrent force would become harder in the coming decade. "The quantitative nature of the problem and the uncertainties" would render the effort to achieve strategic deterrence "extremely difficult," according to his analysis. Of special concern, the Soviet Union would probably develop "a vast increase" in the weight of a potential attack, which could be delivered "with little warning"—"an essentially warningless attack." Wohlstetter provided little comfort to his fellow Americans, stating that from the aggressor's viewpoint, the decision to launch "a surprise thermonuclear attack might not be an irrational or insane act."[109]

It did not take much of a leap to argue, as the Princeton political scientist and later full-time RAND employee William Kaufman did, that massive retaliation had become "neither feasible nor desirable as a policy of deterrence." In contrast to the all-or-nothing character of the Eisenhower administration's policy, Kaufman reasoned that a successful containment strategy would have to demonstrate the nation's "willingness and ability to intervene with great conventional power in the peripheral areas, after the manner of Korea." Only in this way would the United States "have a reasonable chance of forestalling enemy military action."[110] This point acquired additional persuasive power when placed alongside the "domino theory," another pervasive albeit simplistic notion that said when any country in any part of the world faced a communist takeover, its neighbors were in imminent danger of falling as well.

During the late 1950s and early 1960s the work of these scholars informed the evolution of national security policy. Key ideas from deterrence theory infused the classified National Security Council report "Deterrence and Survival in the Nuclear Age"—otherwise known as the Gaither Report after H. Rowan Gaither, the chairman of the committee that produced the report, the chairman of RAND's board of directors, and, as discussed in the next chapter, a major figure in the development of the modern Ford Foundation and its Behavioral Sciences Program. Despite the Gaither Report's top-secret status, its implications reached the press, much to the delight of Eisenhower's critics, including the Massachusetts Democratic senator John F. Kennedy. During his presidential campaign, Kennedy claimed the outgoing administration's reliance on the threat of massive retaliation had allowed the Soviets to make great headway in the competition for allies in the third world. Not only did the Gaither Report inform Kennedy's presidential campaign, but, as indicated by S. M. Amadae's work, that report also "launched RAND's strategy cadre into national prominence."[111]

Soon after moving into the White House, Kennedy delivered a message to Congress outlining the need for new military capabilities to meet the nation's evolving foreign policy challenges. Only two months before, in January of 1961, Soviet leader Nikita Khrushchev had announced his country's support for "national liberation wars," a policy soon adopted by communist China. The previous year, the North Vietnamese government had declared its support for the National Liberation Front, a coalition of forces including communists seeking to overthrow the American-backed South Vietnamese government. And the year before that, Fidel Castro had led the successful communist revolution in Cuba. Against this background, and with insights from the new science of strategy as a guide, Kennedy called for the development of "a greater ability to deal with guerrilla forces, insurrections, and subversion."[112]

At the very same time, social scientists working on these problems attracted greater attention and funding from the military-academic complex. In 1961 Princeton's Center for International Studies hosted a conference on "internal war."[113] The following year, the military's recently created Advanced Research Projects Agency sponsored a symposium on counterinsurgency held at RAND's Washington office.[114] Another 1962 conference hosted by the SORO and sponsored by the army's chief of research and development focused on applying social science research to "the U.S. Army's limited-war mission." At these and many other similar events military personnel came together with social scientists for the purpose of studying "forms of conflict short of all-out nuclear war and general conventional war, with stress on 'wars of subversion and covert aggression' and the 'Cold War.'"[115]

Of all the counterinsurgency studies, Project Camelot emerged as the boldest. This study, designed by the SORO, had an anticipated cost of $4–6 million. Its 1964 planning documents indicated that Camelot would develop a social systems model useful for predicting the course of revolutionary developments in countries around the world and thereby help the American military to suppress such revolutionary activities.[116]

In sum, by the early 1960s not only had the military and its separate branches established an array of organizations and programs supporting social and psychological research. In addition, a number of specific fields with extensive defense funding conformed to the more general scientistic outlook advocated in social science wartime stories and then in the book of ONR-sponsored social research conducted between 1946 and 1951. Though each field of research examined in this section had its own particular trajectory, taken together their stories reveal how a broad pattern of selective military support contributed to the remarkable prominence of scientistic forms of inquiry and the associated promise of powerful technical applications during the Cold War years. In the case of communications studies, an interwar stream of critical scholarship that sought to expose the antidemocratic implications of propaganda and that valued qualitative case studies became marginalized, while a competing and purportedly value-neutral approach steeped in quantitative analysis and designed for technocratic purposes gained enormous momentum. Starting with World War Two and continuing into the Cold War era, the development of the decision sciences, including operations research, systems theory, game theory, and Brodie's science of strategy, promised to place decisions about military resource allocation and strategic doctrine on an analytically sound, nominally apolitical, and putatively rational basis. In late 1950s and early 1960s counterinsurgency studies aimed to reveal the underlying dynamics of revolutionary movements in a way that would facilitate effective interventions.

Furthermore, social scientists with expertise in a variety of national security–related fields acquired unprecedented visibility and influence during the presidency of John F. Kennedy. As contemporary commentators and historical accounts have often noted, Kennedy demonstrated great enthusiasm for working with the nation's best and brightest. This group of intellectuals, including many social scientists from elite universities such as MIT and Harvard, took a keen interest in policy issues, leading one journalist to dub them "action intellectuals." Among them Kennedy's secretary of defense, Robert McNamara, straddled the worlds of academia and politics with special confidence. In his bold effort to establish a more rational and scientific system of military budgeting and decision making, McNamara turned to a cadre of RAND experts from the fields of defense economics and systems analysis.[117] These initiatives, in turn, enhanced the national visibility of social scientists from the relevant fields of

FIGURE 5. President Kennedy meets with Secretary of Defense Robert McNamara. June 19, 1962. *Cecil Stoughton, White House/John Fitzgerald Kennedy Library, Boston.*

study. The military's research budget for the psychological and social sciences rose accordingly as well, climbing to $31 million by 1964.[118] Those adolescents from Samuel Stouffer's story seemed to have finally grown some whiskers.

The Military–Social Science Partnership under Scrutiny

At the time of Kennedy's assassination in late 1963, probably nobody anticipated that the burgeoning military–social science partnership would soon become a subject of widespread scholarly and political debate. Nevertheless, as indicated below, aspects of that partnership did seem problematic from certain points of view. Especially through the mid-1950s, conservative congressional scrutiny raised questions about the value of certain military social research projects. By the early 1960s, emerging criticism from the New Left presented a different sort of challenge by suggesting that research at RAND and elsewhere could have dangerously militaristic implications. As military support for research increased and as this research became more important in shaping the nation's foreign policy and military affairs, the general threat that military patronage posed to the independence and quality of academic scholarship became more salient as well.

Following partisan conflict over the social sciences during the 1940s, some military social research projects attracted conservative congressional scrutiny during the 1950s. One telling episode involved the Refugee Interview Project (RIP), mentioned briefly earlier. This massive study of Soviet society was carried out by Harvard's Russian Research Center and sponsored by the air force's HRRI. In 1953 the Michigan Republican senator Homer Ferguson said it was hard to understand why the air force had allocated money to study the Soviet social system rather than sponsoring work to identify strategically vulnerable Soviet targets. As noted in David Engerman's insightful history of Sovietology in the United States, rifts within the military concerning the practical value of social research also stimulated controversy over RIP. Social research carried out for this study advanced many academic careers, resulted in significant research publications, and thus seemed productive from an academic viewpoint. However, the consequences of congressional and military criticism resulted in a rather unhappy turn of events. Not only did the air force terminate its funding, thereby making the acronym RIP especially appropriate, but also, "after a long hot summer of controversy," the HRRI itself was shut down.[119]

Furthermore, the controversy over RIP occurred at the height of the McCarthy Era, an especially difficult time for social scientists and their patrons. As noted before, in 1953 Senator McCarthy led a critical congressional inquiry of the VOA, and the military's Human Resources Committee indicated that dangerous funding inadequacies reflected widespread legislative scrutiny of the social sciences. Under these unfavorable conditions, military funding dropped precipitously, by roughly one-third from 1953 to 1954, from about $25 million to $16.5 million. Moreover, almost half of the latter amount supported the development of training devices and thus had a rather narrow focus.[120] As discussed in the next chapter, McCarthyite politics also supported a series of anticommunist congressional investigations of private foundations and their social science activities.

In 1958, a few years after the RIP controversy, another round of conservative scrutiny focused on a RAND-sponsored study published under the title *Strategic Surrender*. Carried out by Paul Kecskemeti, a political analyst who later became a counterinsurgency expert, this project aroused the concern of the Georgia Democratic senator Richard Russell, Jr., who presumed that RAND had funded Kecskemeti in order to examine the conditions under which the United States might decide to surrender. Finding this idea abhorrent, Russell, who chaired the powerful Armed Services Committee, put forth an amendment to the committee's supplemental appropriations bill stating that "no part of the funds appropriated in this or any other act shall be used to pay" for any future study that considers how the United States might surrender. Another, slightly earlier episode (discussed in the next chapter) that has some revealing similarities to the controversy over Kecskmeti's study ended with President Eisenhower signing a 1956 bill that outlawed the secret recording of jury deliberations. Eisenhower supported this legislation following a 1955 Senate subcommittee inquiry of a Ford Foundation–sponsored study of the jury system during which another powerful southern Democrat, Mississippi's James Eastland, strongly suggested that the study in question as well as its scholarly researchers threatened the nation's system of government. Against this background, the amendment proposed by Russell had a good chance of attracting political and media attention. Indeed, national newspapers covered the story with catchy headlines, such as "Ike Blows His Top at United States Surrender Article." According to the accompanying story, a White House phone call brought "military business" to a near standstill for about two hours while military officers "frantically launched a searching probe."[121]

Though, as it soon became clear, the hullabaloo raised by Senator Russell rested on a misunderstanding, the result still shined an unflattering light on this military-sponsored social research project. The course of a Senate discussion pointed out that Kecskemeti had used historical case studies of surrender involving the French, the Italians, the Germans, and the Japanese as a basis for critiquing the Allied policy that demanded "unconditional surrender" from defeated enemies in World War Two. In a brief final chapter, Kecskemeti suggested his analysis might be relevant to war in the nuclear age. Thus his study did not consider how the United States might surrender—though one might have reasonably asked if this possibility could be pursued using his analysis.[122] Nevertheless, the Senate proceeded to vote on Russell's proposed amendment, even after the misunderstanding became apparent, with eighty-eight senators in favor, only two opposed, and six abstentions. As reported in the scholarly journal *World Politics*, the nation thus had a new prohibition on the use of congressional appropriations for social research.[123] While the controversies noted above focused on specific military-sponsored social research projects, those controversies helped to keep conservative concerns about the social sciences in the national spotlight during the 1950s.

A different challenge to military social science arose from the New Left. During the 1960s, this loose mix of liberal and more radical groups, including Students for a Democratic Society and other activist organizations on university campuses, crusaded for civil rights, protested against the Vietnam War, and called for the dissolution of the military-industrial-academic complex. New Left leaders often recognized a crucial transformative role for critical intellectuals including social scientists, such as the German émigré sociologist and political theorist Herbert Marcuse, sometimes called the Father of the New Left, and the left-liberal sociologist C. Wright Mills, who died in 1963. Not incidentally, Mills's 1960 "Letter to the New Left" helped to popularize this term.[124] Starting in the early 1960s, Irving Louis Horowitz, another sociologist and a strong admirer of Mills, emerged as one of the nation's most articulate and prolific scholarly critics of the military–social science partnership.

Early in the decade, Horowitz presented a scathing analysis of a group of scholars he called the "new civilian militarists." Writing in the *American Scholar*, a quarterly journal of public affairs and intellectual commentary, Horowitz noted that this group, whose members came mainly from economics and political science, worked at military-funded think tanks and university centers dedicated to the analysis of international relations and war strategy. He singled out for special rebuke a number of figures associated with RAND, including Bernard Brodie and Albert Wohlstetter as well as the physicist Herman Kahn, the political scientist Henry Kissinger, and the economist Thomas Schelling.[125]

Regarding their efforts to place strategic thinking on a rigorous scientific basis through the application of game theory, Horowitz identified a host of disconcerting biases. To begin with, he attacked their reliance on game theory's standard assumption that there were only "two players," namely, the United States and the Soviet Union, and that these players displayed "symmetry in everything from weapons to morals." This assumption, Horowitz charged, left out a realistic appraisal of the role of other important players, including the United Nations as a mediating force but also other countries from England to China that had their own independent aims and interests. He also found game theory's "utilitarian view of human behavior" unrealistic and limiting, leading to an impoverished analysis of the "mutual strengths and weaknesses of the players" and overlooking the importance of "moral factors in decision-making." Horowitz further questioned the supposition of these scientific strategists that the main players would continue to abide by the "rules of the thermonuclear game," rather than deciding to take actions that could result in a nuclear holocaust.[126] Not incidentally, game theorists themselves sometimes worried about hidden biases in their work: "Even though deep analysis may be carried out on a formal model with apparent objectivity, the investigator's biases and misperceptions may be hidden by omissions or protected by the armor of irrelevant mathematics," observed RAND's Martin Shubik.[127]

Though in making the above criticisms Horowitz attacked game theory's "scientific adequacy," just as one might do when evaluating any other field of research, his analysis also presented damning criticisms of the scientistic orientation that characterized so much social science work supported by RAND and other nodes of the defense science establishment. In Horowitz's words, this work had "the appearance of satisfying the historical identification of science with social welfare, while preserving an emotional identification with the neutral and objective image culled from experimental physics." Yet "far from lessening the threat of nuclear warfare," such flawed studies supported "the most dangerous sort of miscalculations about the options, directives and goals of the other player." Suggesting that one could learn some more general lessons from this particular case, Horowitz pointed to the "dangers" of using "mathematical techniques" without a deeper understanding of social science inquiry. He underscored the limitations of conceiving of "rationality" in a way that isolated it from "reason." And he noted the troubles that arose from the supposed scientific imperative to maintain an "absolute separation of 'is' questions from 'ought' problems."[128]

In short, Horowitz's critique from the emerging New Left indicated that rather than producing work that was value neutral, scientifically precise, and practically useful, the civilian militarists were advancing a misleading view of human decision making and international affairs, while also advocating a dangerously militaristic foundation for American national security strategy. In the following years, Horowitz and a growing number of critics in the nation's political and scholarly spheres would propose that such problems had, in fact, become pervasive, with grave consequences for the social sciences, for the nation, and for the foreign peoples who suffered horrific consequences of social science–informed military strategies and programs.[129]

During the late 1950s and early 1960s, substantial funding increases from the military and other national security agencies also raised the specter that large segments of the social sciences would become subordinate to these powerful patrons. Back in the mid-1940s, some concerned liberal scholars including Robert Lynd and Louis Wirth had warned that following research agendas set by the federal patron could encourage the social sciences to develop in antidemocratic directions. In the early 1950s, the pressures on scholars to produce results deemed useful from the military's point of view had mounted. As DOD's Human Resources Committee noted, the military had decided to limit its support for "basic research" to work that promised to "fill in gaps in scientific knowledge of critical importance to military readiness."[130] This chapter's opening quote from Don Price suggests that in those years American scientists were "struggling to reconcile" an older view of science as an "objective and dispassionate search for knowledge" that aimed to further "the welfare of mankind in general" with a newer view that recognized the "necessity of using science as a means for strengthening the military power of the United States." And in his 1961 Farewell Address, Eisenhower issued his famous warning about the degradation of university scholarship due to the power of the federal purse: "The prospect of domination of the nation's scholars by Federal employment, project allocations, and the power of money is ever present—and is gravely to be regarded."[131]

Social scientists themselves did not engage in extensive hand-wringing about the danger posed by patron power, at least not before the mid-1960s. Yet it seems unwarranted to conclude that prior to that time only a small number of scholars worried about this danger. In fact, in their writings about military projects and defense funding a number of social scientists explicitly identified and discussed the threats to the quality of scholarship and to academic independence. In addition, they identified principles that they hoped would keep such threats at bay.

Perhaps the most widely invoked principle was the claim that science thrives best when it is permitted to exercise self-control. Social scientists, much like their natural-science counterparts, often appealed to the corruption of genetics in the Soviet Union and the destruction of vital parts of the German scientific community during the Nazi era to illustrate the awful consequences when state politics and ideology prevented scientists from governing their own affairs. The vital importance of scientific self-governance also received extensive consideration from prominent historians, sociologists, and philosophers of science, as chapter 4 will discuss. Furthermore, scholars identified more specific principles that could help regulate the conduct of military-sponsored research in ways that protected scholarly integrity and independence. For instance, in a report about studies on persuasion and motivation, the communications expert Wilbur Schramm asserted that the DOD recognized that it had to allow social scientists the freedom to pursue basic research, for "in the behavioral field, as in the physical sciences, a comparatively small advance in the understanding of fundamental principles often makes possible a broad and swift advance in practical application."[132] Other scholars found reassurance in the fact that the military relied heavily on universities as sites of research and training and also depended regularly on university scholars for advice. In the case of the ONR, the psychologist John Darley even declared that if either side were to impose on the other, "it was more likely that the [scientific] advisory groups would do the imposing." Thus the concern that "government funding would result in government planning and control" had, at least in his view, resulted in a process that favored scientific control.[133]

Finally, while the demands of classified research and security requirements posed a special threat to the open discussion and evaluation of scholarship, social scientists working on such studies valued arrangements that often allowed them to publish their results in the open literature in some form, after the proper authorities gave them approval to do so. Because RAND became such an important center for many lines of military-sponsored social research, its publication policies serve as a good example. One of twelve policy principles governing RAND's support of social research stated that RAND would encourage "publication of research" as long as the work in question respected the "limits of security requirements." With this stipulation in place, RAND developed a publication series available to the public, starting with a 1951 book by Nathan Leites that analyzed "the rules which Bolsheviks believe to be necessary for effective political conduct."[134]

Still, though discussion of these principles and their implementation helped, at least to some extent, to sustain the hope that patrons would not direct social research in unsavory directions, the strength of this threat only increased over time. At best the dangers to scholarly integrity and independence could be skillfully managed. But as funding from the military and other national security agencies rose during the late 1950s and early 1960s, and as the research, scholars, and institutions they supported acquired greater policy influence, concerns about those dangers had a greater chance of gaining traction. These developments thus helped set the stage for the growth of more widespread criticism of the military–social science partnership during the mid- to late 1960s.

Before the mid-1960s, none of the three issues discussed in this section had gained sufficient attention to inhibit the growth of that partnership. Still, conservative criticisms had helped to bring about the end of the Refugee Interview Project and the closure of the HRRI in 1953, a year that also saw a sharp decline in overall military social science funding. Voicing concerns associated with the nascent New Left, Irving Louis Horowitz charged the civilian militarists with promoting a dangerous view of international affairs and American strategic interests on the basis of inadequate studies dressed up with misleading quantitative precision and a veneer of value neutrality. In addition, though social scientists elaborated a number of reasons for thinking that the threats to scholarly independence and integrity could be kept in check, the increasing importance of military funding and the growing policy influence of military-sponsored research exacerbated the magnitude of those threats. Throughout the Cold War years the overpowering presence of physical scientists in the defense science establishment kept social scientists in a subordinate and often uncomfortable position as well.

In the Shadow of the Natural Scientists

During World War Two and the early Cold War years physical scientists from the Manhattan Project generation had a dominant presence at the highest levels of defense science policy making. In the post-Sputnik period, their dominance became further entrenched through the elaboration of new federal science agencies and advisory apparatus. As late as 1966, the Defense Science Board, created in 1956 as the military's central science advisory group, did not have a single social science representative.[135] Whether social scientists perceived this issue as a problem surely varied significantly from field to field and from individual to individual. Many scholars from all the individual disciplines and from an array of interdisciplinary fields sincerely believed their work had much in common with the natural sciences. In many areas of inquiry such as game theory and systems analysis, to take just two examples, social scientists also saw valuable opportunities to collaborate with their counterparts from mathematics, the life sciences, and the physical sciences. Nevertheless, the reigning scientific hierarchy meant that if social scientists wanted to gain recognition and funding, they typically had to present themselves as junior citizens seeking to emulate the more mature "hard" sciences. Moreover, as indicated below by two landmark reports from the post-Sputnik years, their subordinate status and corresponding pressures to embrace a scientistic outlook, grounded in quantitative precision and technical sophistication and accompanied by the promise of social engineering applications, had no end in sight.

The first report emerged soon after the Soviets launched *Sputnik 1*, the first artificial earth satellite, in October of 1957. Much to the horror of the American people, their archenemy could now hurl intercontinental ballistic missiles tipped with nuclear warheads toward American cities and other targets. In the ensuing panic fueled by the perception that the Soviets were winning the interrelated scientific, technological, and military competitions, the United States redoubled its commitments to the sciences. Among other highlights, President Eisenhower appointed the MIT president and physicist James R. Killian, Jr., to the new position of special assistant to the president for science and technology. Killian then became chairman of the new physical science–dominated President's Science Advisory Committee (PSAC). In the next couple of years, the National Aeronautics and Space Agency (NASA) was established, the military created the Advanced Research Projects Agency (ARPA), and the National Defense Education Act was passed. How could social scientists develop a case for their significance in this context, after the physical sciences had grabbed the spotlight once again?

One possible answer came from an unexpected source, the nation's vice president Richard M. Nixon, who proposed that in the battle against communism, America should not make the mistake "of going overboard in developing or in putting emphasis on scientific and technical training to the exclusion of training in the social sciences and other fields." Unfortunately for the historian, Nixon did not elaborate on what type of social science he had in mind.[136] But regardless of his personal views, Nixon's initiative led to the creation of a new committee charged with producing a report on the state of the social sciences and chaired by James G. Miller, a psychologist who had moved in the mid-1950s from the University of Chicago to the University of Michigan, where he became director of the Mental Health Research Institute. Fifteen well-known scholars signed the resulting report, "National Support for Behavioral Science." Many also had substantial contact with the military and other national security agencies, including the psychologists Raymond Bauer and Donald Marquis, Clyde Kluckhohn from anthropology, Max Millikan from economics, and the sociologist Samuel Stouffer.[137]

According to the Miller Report, the United States enjoyed a clear lead in the social sciences, but the nation also needed to push forward with determination or else risk losing its Cold War advantage. In this picture, American social and behavioral scientists had made remarkable advances by embracing a modern and scientifically rigorous approach. In one telling passage, the Miller Report recognized "an important region of overlap between behavioral science and the humanities, particularly, history, literature, philosophy, and religion." But the report quickly qualified this point by asserting that the behavioral sciences went further, by building on and clarifying the wisdom from the humanities and by attempting to "formulate and test precise laws." Meanwhile, the report also claimed, scholars in the Soviet Union remained hamstrung by "Communist dogma." "Marxist doctrine" had "held back the social sciences," while "Pavlovian doctrines" had "retarded much of Russian neurophysiology and psychology." But the Miller Report also warned that America could succumb to complacency, allowing the enemy to concentrate its resources and thus gain a dangerous advantage in the social sciences. Of special concern, the Miller Report warned that "a breakthrough in the control of the attitudes and beliefs of human beings through exceptionally effective educational techniques, drugs, subliminal stimulation, manipulation of motives, or some as yet unrecognized medium" could become "a weapon of great power in Communist hands." To maintain American superiority and develop "effective countermeasures," the nation therefore needed to redouble its social science commitments.[138]

A reevaluation of defense social science programs inspired a second and much more extensive study whose final report displayed a scientistic outlook just as boldly. In 1957, the same year the first Sputnik was launched, the Defense Department initiated planning studies to examine the military's long-range social science needs. Two years later the DOD issued a contract for carrying out these studies to the Smithsonian Institution, which created a Research Group in Psychology and the Social Sciences. Led by the military psychologist Charles Bray, the Smithsonian's research group had six specialized task groups. Sixty-five scholars served as consultants, including Lyle Lanier and Wilbur Schramm. Defense science leaders remained directly involved with the planning process through a steering group chaired by Dael Wolfle, an experimental psychologist who had worked at the OSRD during World War Two and then became a long-time executive secretary at the American Psychological Association.[139]

In a 1962 article published in the *American Psychologist*, Bray explained the purpose of this major review and claimed the nation's defense leaders needed a "technology of human behavior." This "key concept," Bray continued, implied that just as research in the natural sciences and engineering had improved DOD's "sophistication and inventiveness about the production of physical objects," the social sciences aimed to improve the military's "sophistication and inventiveness about people." Repeating a common complaint made by social scientists, Bray asserted that military leaders still tended to "depend heavily on traditional wisdom and common sense." Thus only seldom did their "decisions about people" rest on "an adequate background of technical knowledge." With this problem in mind, Bray emphasized that the desired "technology of human behavior" would depend on the advance of "scientific theory," including the use of concepts that could be "expressed mathematically." The resulting "technological information" would be "based on controlled observation," preferably "expressed in formulas, tables, and graphs." The general point seemed worth repeating: the purposes of national defense required testable claims and "quantitative information."[140]

Bray added that the social sciences themselves had much to gain, for they needed DOD's ample resources to pursue ambitious work along the desired lines. In fact, Bray claimed that the military, and only the military, could provide sufficient "support for the systematic, long-range study of human behavior." Pursuing such work would surely be "time consuming and expensive." It would require "superb facilities, adequate interdisciplinary and technical help, and continuity of support" for about one hundred psychologists and social scientists. Looking at "the present structure of research support," the Smithsonian's research group had identified the military as the most "logical source" of such support, reported Bray. No other patron could be expected or entrusted to provide the required resources.[141]

While these two reports attested to the growing presence of the social and psychological sciences in national security circles, the behind-the-scenes stories of these reports also underscored the social sciences' problematic second-class status in the physical science–oriented defense science establishment. The Miller Report's harrowing discussion about the need to stay one step ahead of the enemy's efforts in mind control did attract some attention from elite scientists. But the end result proved disheartening, to put it mildly. Following the report's publication, the recently established PSAC expressed interest in discussing these issues with Miller, leading to a meeting that included PSAC's physical scientists, Miller, and General Jimmy Doolittle representing the military. Some years later Miller recalled that the meeting inspired "a lively discussion" along with a "good deal of controversy." Yet any PSAC interest in pursuing the Miller Report's recommendations evaporated after General Doolittle delivered a "stem-winding sermon" criticizing social science.[142]

Since social scientists would rarely have a chance to participate in PSAC's work, one wonders if through that meeting they could have obtained a toehold there. In addition, the Miller Report recommendations had already called for the creation of a behavioral science panel to advise the special assistant to the president for science and technology as well as increased funding for social research, training, and facilities from a combination of private and public sources.[143] But the general's negative view "silenced" the group.[144] Moreover, with no other arrangement in place for pursuing those recommendations, Miller's social science group completed its work with little fanfare.

The second report discussed above gave social scientists greater opportunities to promote their work to the military science establishment, but palpable pressures also kept them mindful of the ever-present need to please their physical science superiors. Since Charles Bray had a key role in producing this report, his unpublished observations about social scientists' subordinate position have particular significance. Recall that Bray's military ties went back to his involvement in OSRD's Applied Psychology Panel during World War Two. He also wrote the panel's official history, wherein he noted that military support for psychology seemed in danger of withering away, just as it had after World War One.[145] Though the military did, in fact, become the largest patron of psychology during the early Cold War years, Bray, who held various positions during the mid-1950s at the Air Force Personnel and Training Research Center (Lackland, Texas), never lost sight of the constant need to defend the social and psychological sciences before the many doubting Thomases in defense science circles, whether they came from the military side, as in the case of General Doolittle, or from the scientific and engineering sides, as in the cases of Vannevar Bush, Karl Compton, and Edward Teller. An experienced and savvy veteran, Bray hoped the Smithsonian research group, with its six working panels and sixty-five scholarly consultants, would produce some good ideas. But he also knew that the future of military social science activities would be determined largely by non–social scientists.

Reading the final reports from the working panels did little to relieve Bray's concerns about satisfying the Defense Department's physicists and engineers. As he explained in an unpublished 1962 note, these reports "left much to be desired as public relations devices" vis-à-vis this group. He therefore decided to write the final report himself, which he submitted in 1960 to the director of defense research and engineering.[146] In this unpublished report, Bray complained that "no single military department appears to have a consistent, continuing interest in the growth of this [social science] technology," at least partly and perhaps largely because "the subject matter is often somewhat peripheral to the major concerns of the natural science, engineering, and medical groups who have responsibility for research and engineering in the military establishment." Bray added that owing to their peripheral status, social scientists and psychologists widely suspected that "military support of their research and development is unstable and not to be trusted."[147]

In an unpublished 1960 letter to Ithiel de Sola Pool, a political scientist at MIT's Center for International Studies and a vigorous advocate of the military–social science partnership, Bray elaborated on those difficulties. Though social science research could contribute to more effective policies, Bray believed that the average policy maker typically had "only the general social science background common to our culture," and even if he possessed the necessary education "to read and understand," he had "no time to read the relatively long discussions" contained in the reports of research results. These results could still influence his thinking "either indirectly through the general culture or through advisors who are supposed to do his reading for him." But as Bray saw it, the channels for ensuring that social science research received adequate attention remained rather inadequate. Whereas policy makers regularly turned for advice to natural scientists and engineers employed as staff members, and "a whole mechanism in the Defense R and D establishment" ensured that policy makers obtained "qualified advice on natural science questions," the social sciences had "nothing comparable." Furthermore, natural scientists and engineers who had extensive control over military research funding tended to regard social scientists "as historical and literary folk who do not have a technique for making controlled observations and testing the relatively subjective opinions which arose from the historical-literary approach." How then, Bray asked in his letter to Pool, could social scientists convince natural scientists and engineers that "we are increasingly scientific and technological?"[148] Despite Bray's deep worries as revealed in these unpublished documents, he presented a rather positive outlook in his 1962 published report in the APA's flagship journal, even going so far as to declare that the air force, navy, and army "ably" provided support "across the whole broad field of psychology and the social sciences."[149]

In short, social scientists eager to work with the defense science establishment continued to face formidable obstacles owing to their subordinate position vis-à-vis the nation's physical science elite and the lack of consistent support from their military superiors. As indicated by the stories of the Miller Report and the report for the DOD, those difficulties remained as strong in the post-Sputnik years as they had been ever since World War Two. These two cases also underscore the need for historians to look behind those and other published reports. Whereas the published materials regularly emphasized the deepening scientific maturity and the growing military value of the social sciences, unpublished sources offer a more complicated picture that highlights their institutionalized marginalization in the natural-science-dominated defense science establishment and reveals how that marginalization made scientific boundary work crucial to the military–social science partnership throughout the first couple of Cold War decades.

Conclusion

This chapter has presented a complex story about military patronage and its significance for the social and psychological sciences in Cold War America. In the early postwar years, the future of military funding for these sciences seemed quite uncertain. Moreover, social scientists and their allies who were eager to establish a robust partnership with the military faced a number of difficulties rooted in major political and science policy developments of those years. As indicated in the previous chapter and as developed more fully in the present one, conservative politicians regularly attacked the social sciences and a wide array of federal agencies that worked with them, including agencies concerned with national security issues, such as the OWI and the OSS during World War Two, and the Voice of America during the Cold War years. The establishment of a massive governmental system for conducting background checks on federal employees, together with aggressive programs launched by the executive and legislative branches to expose un-American activities, meant that any agency supporting the social sciences would probably come under anticommunist scrutiny. The dominance of natural scientists and especially physical scientists in the massive Cold War defense science establishment also continued to generate questions about the scientific status and social implications of the social sciences. Speaking at the very first meeting of the military's Human Resources Committee, Vannevar Bush suggested not only that the committee's work had an unusual, vague, and experimental nature but also that the social sciences included some kooks. These conditions ensured that within defense science circles, social scientists often found themselves on the defensive. Indeed, throughout the late 1940s and early 1950s, Human Resources Committee documents complained about a serious lack of respect and a corresponding dearth of adequate military funding.

But prominent social scientists also worked hard to build a solid case for greater funding and influence. As revealed by an examination of the war stories written by influential scholars such as Paul Lazarsfeld, Charles Bray, and Alexander Leighton as well as by the successful journalist Stuart Chase, social scientists tried to move from defense to offense, as they championed a viewpoint that had much in common with the scientistic accounts presented by Talcott Parsons and other SSRC representatives during the NSF debate. In the following years, influential scholars from all the major disciplines repeatedly invoked crucial differences between scientific social inquiry on the one hand and humanistic studies, ideology, and undisciplined speculation on the other. Just in case anybody had missed the message, the 1958 Miller Report and the more extensive 1960 report on military social science activities written up by Charles Bray hit the scientistic notes loudly once again.

Furthermore, extensive funding from the DOD and from the individual military branches reinforced those scientistic commitments during the Cold War years. Examination of three lines of inquiry that attracted substantial military support—communications studies, the decision sciences, and the science of strategy—has indicated a pervasive emphasis on quantitative and predictive modes of analysis that promised to yield objective, instrumentally useful, and allegedly value-neutral knowledge. Much of this research was carried out by scholars at leading institutions of higher education, at new interdisciplinary research centers, and at new military-funded organizations including the Rand Corporation and the SORO. By the early 1960s, military programs and policies supporting such work had become much more robust than they had been in the immediate postwar years. Although SSRC scholars had fervently argued during the postwar NSF debate that the social sciences had a great deal in common with the natural sciences, critics from the natural-science and political communities had managed to remove from serious consideration the bid to include the social sciences in the proposed agency, as noted in chapter 1. Against this background, military funding for scientistic work accompanied by the anticipation of powerful social engineering applications has special importance. Moreover, though defense funding naturally carried the expectation that such research would contribute to strengthening American military power, many lines of military-funded research also acquired scholarly significance far beyond their military relevance.

Yet the position of the social sciences within the military science establishment also continued to generate questions about their scientific standing and practical worth. During the 1950s, some military-supported social research projects attracted conservative criticism, including the RAND study on wartime surrender and the air force–funded Refugee Interview Project. These episodes fit into a longer tradition in which conservative legislators charged that social science research threatened the national interest and American values. From a different angle, analysis from the New Left as presented by Irving Horowitz claimed that the work of the civilian militarists was, in fact, neither scientifically sound nor value neutral. Nor did Horowitz find such work helpful from a strategic viewpoint. Quite the contrary, his analysis suggested that their work made the possibility of nuclear war more likely. As defense funding increased over the years and as military-sponsored research acquired significant scholarly and strategic influence, the more general threat to scholarly independence increased in importance as well. Last but certainly not least, within the defense science establishment, the social sciences remained far below the natural sciences in terms of their status, funding, and influence. Residing on the margins of the defense science establishment, leaders of military social science programs continued to struggle with the need to impress especially the physical scientists of the Manhattan Project generation, as indicated so vividly in the behind-the-scenes stories of the Miller Report and the study on military social science activities written up by Charles Bray.

It would be difficult to overestimate the importance of the military in the Cold War patronage system. Yet social scientists also sought and succeeded in acquiring significant support from other powerful patrons. This is not simply a matter of historical fact. According to a widely held belief at the time, if any science depended too heavily on a single patron, that patron could exert an unseemly degree of control over the direction of scientific research and thereby hinder its healthy development. Furthermore, science policy discussions regularly suggested that private patrons enjoyed a greater degree of independence from political pressures than the military and other federal agencies. At the very beginning of the nuclear age nobody could have foreseen that the Ford Foundation would have much importance for the social sciences. But by midcentury Ford had become the largest philanthropy in the world and the single most important private patron of academic social science, or what Ford's leaders often called the behavioral sciences. We have seen that this term also appeared in military documents. As the next chapter will explain, this common usage had deep implications for the Ford Foundation's position within the emerging patronage system and for corresponding efforts by social scientists to establish the scientific character and practical value of their work.

3

Vision, Analysis, or Subversion?

The Rocky Story of the Behavioral Sciences at the Ford Foundation

In sum . . . these differences spell a revolutionary change . . . the field has become technical and quantitative . . . specialized and institutionalized, "modernized" . . . in short, Americanized. Twenty-five years ago and earlier, prominent writers, as part of their general concern with the nature and functioning of society, [studied man] in broad historical, theoretical, and philosophical terms and wrote treatises. Today, teams of technicians do research projects on specific subjects and report findings. Twenty-five years ago [the field] was part of scholarship; today it is part of science.

–Bernard Berelson, Ford Foundation Behavioral Sciences Program director (1951–1957), 1964

The private foundations have "promoted a great excess of empirical research . . . 'scientism' or fake science . . . to the detriment of our basic moral, religious, and governmental principles."

–Report of theReece Committee directed by Republican representative B. Carroll Reece, 1954

As postwar federal funding for social science became a hotly contested issue, leaders at the large private foundations speculated that philanthropic support could make a vital contribution. Writing in 1949, former Rockefeller Foundation vice president Edwin R. Embree explained that many subjects that foundations had focused on for nearly half a century, including the natural sciences and medicine, were already well established and had extensive public patronage. Thus the foundations needed to pursue other opportunities, including what Embree called the "heroic development of the human studies." Equally important, he argued that private support provided a crucial alternative or complement to public patronage, as a private foundation was less vulnerable than a public agency to political influences that might direct funding in unhealthy ways.[1] In the postwar years, many respected voices in the academic, philanthropic, and political arenas agreed with Embree's position. Indeed, during the postwar NSF debate common rhetoric about a new science *foundation* reflected a widely shared understanding that the privately governed, philanthropic foundation would be less susceptible to manipulation by political interests than a public agency and thus would be better able to provide risk capital for controversial but valuable projects. Also, as discussed in chapter 1, the worry that including the social sciences would invite political interference with NSF's activities constituted a central reason for keeping them out of the proposed agency.

Still, what kind of social research deserved philanthropic support? And how could foundations allocate their resources most effectively in advancing the social sciences? According to an NSF study, between 1936 and 1951, forty-four private foundations had allocated 35 percent of their research support for the social sciences, compared with 25 percent for the medical sciences, 15 percent for the biological sciences, 10 percent for the humanities, 10 percent for interdisciplinary and other research, and 5 percent for the physical sciences.[2] Nevertheless, as the first section of this chapter will explain, leaders from the large foundations, including those associated with the massive Carnegie and Rockefeller fortunes, experienced much frustration and confusion when it came to questions about the social sciences' scientific status and practical value. In his 1952 history of the Rockefeller Foundation, the organization's former president, Raymond Fosdick, announced his worries in blunt fashion at the outset, in the table of contents. Many chapters had neutral-sounding titles such as "Medical Research and Psychiatry," "The Natural Sciences," "Experimental Biology," "Tools of Research," and "Agriculture," but one chapter title referred to "The Problem of the Social Sciences."[3] In this chapter, Fosdick claimed that in the process of advancing their technical skills, social scientists diverted their energies away from the great moral challenges of the times: "Beyond the questions of social fact lie the questions of social value, of morals and ethics. With the problems of mankind calling for perspective and vision, our social scientists cannot be merely analyzers and computers."[4]

In the broader history of philanthropic funding, the modern Ford Foundation and its extensive involvement with the social sciences have enormous importance. The historian of science Hunter Crowther-Heyck has proposed that Ford was "easily the most influential single patron of the behavioral and social sciences in the 1950s."[5] In a similar vein, Roger Geiger, a leading historian of American higher education, writes that Ford "became the most significant external arbiter of the development of university research outside of the natural sciences" from the mid-1950s to early 1960s. Following the recommendations of a 1949 study directed by the chairman of RAND, H. Rowan Gaither, the modern Ford Foundation created five new program areas. Between 1952 and 1962 these programs granted $82 million for social or behavioral science research. Program I, which had a mandate to promote world peace, supported research on international relations. Program II, which aimed to advance democracy, funded work in political science. And Program III, which concentrated on economic progress, supported the fields of economics and business administration. Those three programs combined to provide about $44 million, but Program V, the Behavioral Sciences Program (BSP), became the one "most intimately involved with university social science," as Geiger also notes.[6]

After discussing the twists and turns in philanthropic funding for the social sciences before the advent of the modern Ford Foundation, this chapter examines the origins of the BSP, its strategy for advancing the behavioral sciences along scientistic lines, and the political and institutional pressures that severely compromised BSP's efforts. The BSP's mission focused on strengthening the behavioral sciences in a manner that drew heavily on the understanding of science presented in Vannevar Bush's *SEF* and that had much in common with military social science programs. Under the leadership of Bernard Berelson, the program's first and only director, the BSP initially emphasized the development of scholarly resources, based on the premise that objective, basic research would provide the basis for realizing the more practically oriented social, economic, and political objectives pursued by the foundation's other programs. But Berelson's program encountered serious difficulties. Despite the alleged insulation of the private foundations from political pressures, Ford and the other foundations faced a series of anticommunist congressional investigations that examined whether their social science activities supported left-wing agendas. In addition, support among the foundation's leaders faltered, partly because BSP's scholarly orientation did not excite them and partly because it did not seem to be worth the public relations risk, especially after one BSP-funded study of the jury system came under harsh congressional scrutiny.

Under pressure from this mix of external hostility and internal displeasure, the BSP, in a critical turn of events that has received little attention, shifted its focus to behavioral science applications by the mid-1950s. As my analysis indicates, this applied tilt made the BSP more like Ford's other programs that sought to establish close links between scholarly social science, practical problem-solving, and public policy initiatives both at home and abroad. Furthermore, an examination of Ford-funded research on juvenile delinquency and development studies indicates that the foundation's early plans to use the BSP for promoting basic studies with an ostensibly apolitical and objective outlook gave way to a messier situation in which the distinctions between social research and social reform became blurry. Indeed, though in 1957 the foundation decided to close the BSP altogether, it continued to provide extensive funding for research that combined scientific analysis with advocacy for particular social values and political objectives.

From Troubled Past to Postwar Confusion

In the early twentieth century, private foundations played a starring role in the growth of higher education and scientific research. Regarding their engagement with the social sciences, much of the best historical literature has emphasized that the large private philanthropies promoted a scientistic model during the interwar period. The Rockefeller philanthropies and the Carnegie Corporation in particular insisted on a distinction between research and reform, encouraged the development of quantitative methods, empirical inquiry, and the systematic collection of data, and claimed a commitment to nonpartisanship and ideological neutrality.[7] But another part of the story of philanthropic funding is just as significant, though not so well known. Namely, the development of philanthropic programs for the social sciences experienced many twists and turns from the late 1910s to the early post–World War Two years. Debate about the plausibility and the value of a scientistic model along with its associated value-neutral, disinterested investigative stance persisted, stimulating significant controversy both inside and outside the Carnegie and Rockefeller foundations.[8] Consideration of this complex story will thus provide a crucial background for understanding how the Ford Foundation's BSP contributed to the intertwined and long-standing debates over scientific identify, social purpose, and political acceptability.

One of the first social science projects undertaken by the new, modern, and scientifically inclined philanthropic foundations backfired badly. In 1914, the Colorado State militia, which had been called in to keep the peace during a labor strike at the Rockefeller-controlled Colorado Fuel and Iron Company, opened fire at striking workers and their families, resulting in some forty deaths. Following the "Ludlow Massacre," the Rockefeller Foundation (f. 1913) asked Harvard-trained political scientist and former Canadian labor minister William Lyon Mackenzie King to study the explosive problems of industrial relations. But the hiring of King, who was also a personal adviser to the Rockefeller family, provoked public outcry and strengthened anti-Rockefeller hostility. A national investigation soon commenced.[9]

Conducted by the Industrial Relations Commission and led by Progressive lawyer Frank Walsh, the investigation culminated in a biting 1916 report. Among other criticisms, it called the Rockefeller Foundation "a menace to the national welfare" and recommended further investigation of the foundations. Walsh himself proposed that the Rockefeller Foundation should be closed, its money confiscated and used to improve the conditions of workers.[10]

For the next few years, the future of Rockefeller social science support appeared grim as the young foundation's architects pursued initiatives in more reputable scientific areas, especially medicine and public health. Yet following the First World War, the Rockefeller philanthropies recognized that the terrible problems of human behavior and social relations were too important to ignore. During the 1920s, the Laura Spelman Rockefeller Memorial (LSRM), one of many foundations created with the Rockefeller family's enormous wealth, took the lead in navigating those turbulent waters.

Led by experimental psychologist Beardsley Ruml, the LSRM developed the country's most extensive program of extra-university social science patronage while pursuing a scientistic strategy that promised to place Rockefeller funding beyond political controversy. Under Ruml, the LSRM indicated its interest in nonpartisan and value-neutral empirical scholarship as well as theoretical work that could be tested with quantitative data gathered, whenever possible, through systematic methods. In general, the LSRM aimed to make social science inquiry equivalent in rigor to natural science inquiry. Instead of advancing social welfare through "legislation" and "social, economic, or political reform," Ruml's organization looked to support "men of competence" who pursued social research "in a spirit of objectivity." The LSRM also provided major funding for institutions to support topflight social science programs of research and higher education in the United States and Europe.[11]

During the 1920s, the Carnegie Corporation adopted a similar nonpartisan strategy. For example, the corporation helped establish an Institute of Economics in Washington, D.C., the forerunner of the Brookings Institution, for "the sole object of ascertaining the facts and of interpreting these facts to the people of the United States" and "in the interest of the common good"—at least so stated the corporation's *1922 Annual Report*. From this perspective, the disinterested professional social scientist would advance the public welfare by discovering and putting forth the objective facts of human behavior and social life, permitting the average citizen as well as more powerful private and public decision makers to act in an informed and rational manner.[12]

But within a decade Ruml's strategy of building up social science personnel, methods, and institutions lost much of its appeal in the context of a devastating economic crash and political upheavals in Europe and Asia. After 1929, Rockefeller philanthropic leaders closed the LSRM and relocated their major social science effort within the Rockefeller Foundation's own Division of Social Sciences. By the mid-1930s, the foundation's emphasis had shifted from "general institutional research in the social sciences" to "specific areas of activity" that promised to aid "in the solution of pressing social problems." Ruml had always believed the social sciences would have practical value. However, the Rockefeller Foundation now promoted tighter links between research and action as it focused on the fields of international relations, social security, and public administration.[13]

Furthermore, as many influential American and European thinkers became increasingly disenchanted with the ideal of disinterested, nonpartisan scholarship during the 1930s, the private foundations themselves became engaged with this issue. From 1933 to 1945 dozens of European and mainly Jewish social science refugees supported by the Rockefeller Foundation found jobs in American universities, with the largest number receiving appointments at New York's New School for Social Research. Though Rockefeller Foundation leaders believed they were helping empirically minded European scholars who had rejected the continental predilection for metaphysical speculation and philosophical musing, the New School's director, Alvin Johnson, hoped these scholars would "retain the values" of their "European university discipline," where "philosophy lay at the center" and extended to "the state and society." A value-oriented approach also appealed to the New School's founders, including the philosopher-educator John Dewey and the historian–political scientist Charles Beard. Critiques of value neutrality and the apolitical stance found further expression in the Rockefeller-sponsored *Encyclopedia of the Social Sciences*, a monumental, fifteen-volume landmark of interwar scholarship for which Johnson served as associate editor.[14]

The effort to reconsider the course of the social sciences deepened within the large private foundations as well. As he left the Rockefeller Foundation in the late 1930s, Social Science Division leader Edmund Day, an economist, wondered "what is it that we have in mind when we so glibly refer to scientific work in the social field?" Joseph Willits, another economist who replaced Day, also questioned the previous enthusiasm for nonpartisanship and a corresponding determination that scholars should separate political and moral concerns from social scientific inquiry. While Willits thought such an orientation might be appropriate for certain areas of inquiry, including parts of economics, he found the effort to banish value-oriented studies altogether unwise, a point we will soon return to. Over at the Carnegie Corporation, president Frederick Keppel reached a similar conclusion, as he called the private foundations "the chief offenders in forcing the techniques of research which developed in the natural sciences, where experimentation is relatively simple, where verification is usually possible, where controls are available, into the social sciences," where he supposed "conditions are very different."[15] In short, on the eve of World War Two, an effort to reconsider the intellectual foundations and social purposes of the social sciences had advanced inside the major foundations.

During the war, foundation personnel and foundation-supported scholars used their expertise to fight the Axis powers. Carnegie Corporation board member Frederick Osborn became director of the War Department's Information and Education Division, which commissioned the famous *American Soldier* study led by Samuel Stouffer. Both the Rockefeller and Carnegie foundations also relied on communications experts to redirect American public opinion away from isolationism in favor of recognizing the nation's international responsibilities. Following the war, many trustees, officers, and social scientists took pride in their wartime activities.[16]

Yet in the immediate postwar years, their wartime experiences did not dictate any particular course of action. The postwar NSF debate made the state of uncertainty about the social sciences' scientific identity and social value painfully problematic. Within the private foundations, the concerns of the prewar years reappeared, and confusion about how to proceed ran rampant. Given the huge postwar expansion in federal science funding, private patrons saw they could no longer compete with the federal government for influence. "Foundations, as now appears likely, are to contribute less than 1% of the future national research budget instead of perhaps 40 or 50% as formerly," noted the Carnegie Corporation *1948 Annual Report*.[17] It was in this context that former Rockefeller Foundation vice president Edwin Embree proposed that private patrons had a great opportunity to support the "heroic development of the human studies." However, his proposal left the sticky problem of deciding what to support in this contested area unresolved.

One approach would be to take wartime work as the model, cultivating a practically oriented and predictive social science, grounded in research organized about real world problems and pursued through carefully controlled empirical studies, with an emphasis on testing specific hypotheses and using, whenever possible, quantitative techniques of analysis. Charles Dollard, a researcher from the *American Soldier* study who became Carnegie Corporation president in 1948, observed with a mixture of concern but also favor that natural scientists were demanding that social scientists become real "scientists . . . dealing in fundamental theory which can be tested, and struggling to achieve laws and generalizations . . . to predict what men or groups of men will do or not do under stated conditions."[18] According to this viewpoint, social scientists needed to embrace a universally valid scientific approach that would provide instrumentally useful knowledge, thereby allowing leaders from other sectors of society, such as politics and business, to make better decisions and take more effective action.

Since many foundation leaders had worked on military and intelligence projects during the war and then retained close ties to national security agencies after the war, the practically oriented, instrumentalist, and purportedly ideologically neutral approach had many supporters. Just looking at scholars from the *American Soldier* study, one finds Donald Young, who became SSRC president briefly and then Russell Sage Foundation president; Leonard Cottrell, also at Russell Sage; Leland Devinney, the assistant director of the Rockefeller Foundation's Social Science Division; and Dollard, who became a RAND trustee, the chairman of the Panel on Human Relations within the military's Committee on Human Resources (CHR), and a member of the air force Human Resources Research Institute's (HRRI) Morale Advisory Research Council. All the above scholar-administrators were sociologists, though representatives from other social science disciplines had equally important positions elsewhere. The political scientist Pendleton Herring was SSRC's president for twenty years, a member of HRRI's Morale Advisory Research Council, and a CHR consultant. Another political scientist, Dean Rusk, was a Rockefeller Foundation trustee and president, secretary of state for Far Eastern affairs, and then secretary of state in the Kennedy administration. The psychologist John Gardner served as a Carnegie Corporation trustee and president and also as a CHR consultant.

Yet, as discussed before, a number of well-known left-liberal scholars found that view of the social sciences simplistic and inadequate. These critics of scientism included the Swedish economist Gunnar Myrdal, who carried out the monumental study of U.S. race relations sponsored by the Carnegie Corporation, as well as the sociologists Robert Lynd and Louis Wirth, who had both received Rockefeller Foundation support and also worked closely with the Rockefeller-supported SSRC. During the 1930s and continuing in the 1940s, these scholars argued that social research modeled after the natural sciences could not provide the ethical guidance and critical analysis that national leaders and American citizens desperately needed.

Inside the Rockefeller Foundation humanistic critics also challenged the scientistic approach, in this case for disregarding humanistic values and fixed moral principles. In the early Cold War years, extensive discussions among the president, trustees, officers, and academic consultants focused on reevaluating the foundation's social science program in light of frightening developments. Participants pointed to the devastation wrought by World War Two, the deepening conflict with the Soviet Union, the communist threat to the "American way of life," and the possibility of a third world war. According to Harvard emeritus professor of philosophy William Ernest Hocking, the inappropriate extension of the "scientific spirit" and quantitative methods into the human realm had encouraged "a general sag in the moral texture of the country." As explained in a marvelous book by historian Peter Novick, such critics of moral relativism feared "the 'neuter' qualities of objectivity and detachment . . . had morally disarmed the West for the battle against totalitarianism."[19] A special foundation committee led by Chester I. Barnard, a trustee who became the foundation's president from 1948 to 1952, confirmed that the world was suffering from a moral crisis and thus recommended a major new investment to support studies that aimed to provide moral guidance.[20]

But as the discussion continued, serious differences of opinion combined with political pressures to limit the anticipated support for research with an explicitly moral dimension. As noted before, Raymond Fosdick proposed that social scientists should go beyond the study of social facts to consider questions about social values, morals and ethics. In line with Fosdick's concerns, the Rockefeller Foundation did provide some money for value-oriented studies, for example, through a new emphasis on legal and political philosophy as well as a new Program of Advanced Religious Studies at New York's Union Theological Seminary.[21] Recent work on the history of international relations theory shows that the Rockefeller Foundation also provided vital support to Hans Morgenthau, Reinhold Niebuhr, Paul Nitze, and other prominent intellectuals who rejected the effort to turn the study and practice of politics into a "science."[22] In the early 1950s, Willits proposed allocating 10 percent of his Social Science Division's budget to philosophical and ethical studies as well. But owing to conflicts with his superiors and McCarthyite pressures, his modest 10 percent proposal failed.[23] For similar reasons, an effort to create a new foundation division of morals and ethics collapsed. Thus, in the end, general agreement on a worldwide moral crisis did not lead the foundation to a coherent social science plan.

During the same years, discussions within the Carnegie Corporation yielded strong differences of opinion over how to proceed as well. In 1948 nearly 75 percent of Carnegie grants went to the social sciences, compared with less than 30 percent just two years earlier.[24] But in contrast to Dollard's enthusiasm for an instrumentalist and predictive social *science*, Russell C. Leffingwell, Carnegie's senior trustee and conservative corporate lawyer, preferred the term "social studies." In his view, "economics" was "really not a science" but "the study of political economy," and he asserted that the entire field of study had a "political" and "controversial" character. Reflecting on this schism within the organization, Dollard later recalled fighting battles with some trustees who "didn't understand what the social sciences were about." With opinion so divided, Carnegie ended up supporting various social science projects, but more on an "ad hoc" and "opportunistic" basis than as part of an integrated plan.[25]

In the early post–World War Two years, then, both large foundations continued to provide important resources for social science, often in close cooperation with military and intelligence programs, in fields such as international affairs, national security, area studies, and communications research. However, inside and outside the foundations widespread debate about how to proceed inhibited the development of a more coherent program for advancing the social sciences. Basic questions about what exactly scientific work in the social field meant had been raised during the interwar era, received blunt expression in Edmund Day's remark, and persisted into the postwar years. Whether and how social scientists might pursue wisdom, vision, and ethical inquiry had arisen during the 1930s and then gained traction again after World War Two, as revealed by the divisive postwar planning discussions and less-than-rosy assessments of prior efforts by Carnegie and Rockefeller insiders. In addition, as Roger Geiger has noted, these two foundations scaled back their social science commitments at midcentury because of the bold plans launched by another foundation, the recently enlarged and indeed humongous Ford Foundation.[26]

The Promise of the Behavioral Sciences

The origins of the Ford Foundation's interest in the behavioral sciences lie in a landmark document titled *Report of the Study for the Ford Foundation on Policy and Program* (hereafter, 1949 *Study Report*), produced under the leadership of H. Rowan Gaither. In setting forth the basic framework for Ford's future activities, this report and subsequent BSP planning documents presented the behavioral sciences in ways that reflected wider concerns about their scientific legitimacy and social utility while also articulating a scientistic approach as the basis for Ford's work in this area. Remarkably, this vision of the behavioral sciences included no hint that the challenges to scientism raised in so many places during the 1930s and then again in postwar policy discussions at Rockefeller and Carnegie might be legitimate. The main figures involved with running the BSP found the scientistic approach, along with its instrumentalist and technocratic implications, appealing as well. These figures included Gaither, Donald Marquis, Hans Speier, and BSP's director Bernard Berelson, all of whom had extensive ties to national security agencies and sought to make social science knowledge available to political and military decision makers. On this point, my analysis supports recent studies on the behavioral sciences that have emphasized the role of funding from both government and private foundations as well as the importance of overlapping academic, philanthropic, and government networks of scholars and program managers.[27]

Originally a family-run, parochial affair with a charter from the state of Michigan, the Ford Foundation (f. 1936) underwent a dramatic postwar expansion that set the stage for Gaither's report. Until 1945, the foundation's governing board included only Henry Ford and two Ford Motor Company employees. For the first dozen years, the foundation's annual expenditures never surpassed more than about $1 million. But as the beneficiary of nonvoting stock in Ford Motor, the foundation possessed assets that soon soared, making it the world's richest philanthropic organization. As of 1951 it held stock worth $417 million, dwarfing the financial assets of the Rockefeller Foundation at $122 million and the Carnegie Corporation at $170 million. The foundation's board of trustees also added new members from the nation's commercial, political, and academic

FIGURE 6. H. Rowan Gaither, 1955. *AP Images.*

elites. To help the transformed foundation chart a new course, its president, Henry Ford II, turned to Gaither, a San Francisco lawyer.[28]

Besides his legal expertise, Gaither had close ties to the nation's defense science establishment. During World War Two he worked as an administrator in MIT's Radiation Laboratory and served as a liaison between the scientific and military communities. At the Rad Lab he became friends with a number of physicists including Karl Compton, who, as one of Ford's new trustees, had recommended Gaither to Henry Ford II. In addition, Gaither served as board chairman at the new Rand Corporation.[29] After the Soviet *Sputnik 1* shook the world in 1957, Gaither, as mentioned in chapter 2, also became the chairman of a National Security Council–appointed committee that produced the influential classified report "Deterrence and Survival in the Nuclear Age."

To carry out his task as the Ford Foundation's study director, Gaither assembled a study team and staff that included representatives from business and industry but was dominated by academic personnel from education, medicine, the natural sciences, the humanities, and the social sciences. The study team's social science representatives included the psychologist Donald Marquis and the political scientist Peter Odegard, while the staff had two more political scientists, Don K. Price and William McPeak. The resulting 1949 *Study Report*, written by Gaither and his staff, outlined a set of domestic and foreign policy concerns framed in the charged Cold War context as the basis for the foundation's future activities.[30]

As the 1949 report put it, the United States, as the champion of democracy and "free peoples everywhere," faced a life-or-death struggle with the authoritarian, totalitarian powers. Reasoning that the most "critical problems" facing civilization were "social rather than physical in character," the report encouraged the Ford Foundation to use its massive resources to strengthen the forces of democracy. Gaither's team then identified five program areas as those offering the "greatest opportunity for the advancement of human welfare": (1) establishing world peace, (2) strengthening democracy, (3) strengthening the national economy, (4) fostering education in a democratic society, and (5) promoting scientific knowledge of individual behavior and human relations. This last area became the basis for the BSP.[31]

But what were the "behavioral sciences"? In contrast to behaviorist psychology, whose leading figures John Watson and B. F. Skinner emphasized the experimental study of observable behavior in relationship to environmental stimuli, the 1949 *Study Report* rejected an approach that ruled out consideration of mental states, beliefs, attitudes, goals, or emotions. In the late 1940s and early 1950s, an interdisciplinary Behavioral Sciences Study Group located at the University of Chicago and led by the psychologist James Miller set out to develop a universal theory of behavior rooted in general systems theory. But neither did the 1949 *Study Report* advocate this approach. Taking a wide view, one finds the meaning of the term behavioral science(s) could vary significantly, depending on the relevant disciplines, sets of practitioners, subject matters, methods, aims, some combination of the above, or even in other ways.[32] Nevertheless, the 1949 *Study Report*, other Ford Foundation documents, and Gaither's own views established a fairly consistent outlook.

To begin with, Ford's effort supposed that the development of a universally valid scientific methodology and epistemology, but not necessarily the development of a universal theory, would be crucial to success. As the 1949 *Study Report* put it, the behavioral sciences needed to advance beyond the sort of "polemical, speculative, and pre-scientific" inquiries characteristic of much earlier work. To encourage this vital move from the prescientific to the scientific, the Ford Foundation would have to respect "the established similarity of scientific methods throughout both the natural and behavioral sciences."[33] Gaither added that the behavioral sciences exhibited "a startling disorder" and still lacked "the important characteristics of science." Though "a natural science like physics consists of experiment, the accumulation of data, the framing of general theories, attempts to verify the theories, and prediction," Gaither saw "very little of this" in the study of human affairs.[34]

At Ford, the behavioral sciences' promise also reflected the widespread postwar enthusiasm for interdisciplinary studies. Though in the social sciences "the major disciplinary tributaries should flow together to stimulate each other, all too often they moved in isolated streams," complained Gaither. A 1951 BSP planning document confirmed the need to cultivate "interdisciplinary collaboration," "integration," and "consolidation" in various ways: by enabling scholars in "like" disciplines, such as sociology and social psychology, to work on common problems; by encouraging those from "unlike" disciplines, such as "psychology and economics," to work together; by facilitating a "more efficient sharing of the methods" across the behavioral sciences; and by stimulating efforts to overcome a "serious breach" between "theoreticians and empiricists." This document also mentioned the need to "repair" the "break" between "two fundamental approaches to the study of man, the humanistic and the scientific." However, any attention to the humanistic approach would pale in comparison with BSP's overwhelming emphasis on advancing the "scientific" approach.[35]

With these considerations in view, the 1949 *Study Report* presented "a long-range plan" for reform. In an argument similar to the one Vannevar Bush advanced for focusing on "basic" research in *SEF*, the report claimed scientific knowledge accumulates bit by bit, "no one of which is necessarily of practical use until fitted into combination with other bits" that initially appear "to be equally useless fragments." Therefore Ford's effort to advance knowledge per se would be important by itself, a point with special relevance because, as we have seen in chapter 2 and as noted here in the report, "major patrons" of science, especially defense and intelligence agencies, generally favored "other fields and not basic work." Accordingly, the BSP should concentrate on basic studies, building up the "storehouse of verified knowledge of human behavior," while leaving aside "applied or developmental research" for later attention. Because basic research in the behavioral sciences suffered from "deficits" in skilled researchers, the BSP should also focus on training "additional research scientists."[36]

At the same time, however, the 1949 *Study Report* anticipated that the behavioral sciences would play a major role in addressing problems in American society, democracy, and civilization more globally. Indeed, BSP's work seemed crucial to the success of Ford's other major program areas: "permanent progress toward the solution" of the basic problems confronting civilization, "from war to individual adjustment," required "a better understanding of man himself." The report thus proposed that the BSP should establish an adequate base of scientific knowledge relevant to central issues of human welfare—peace, democracy, economic growth, and education—but these issues would be addressed from a more practical standpoint by the foundation's other four program areas.[37]

In discussing the practical uses of the behavioral sciences, the 1949 *Study Report* also identified a privileged position for leaders in government, planning, and the professions. These leaders had a central role in Gaither's conception of a "rationally managed society . . . predicated on the authority of scientific expertise," explains S. M. Amadae in a splendid history of rational choice liberalism.[38]

In the words of the report, greater scientific knowledge would establish "a rational basis for planning and responsible decision making" in an increasingly complex and centralized society.[39] However, various professional groups such as psychiatry and social work generally lacked a strong scientific foundation, the report added. Furthermore, social scientists themselves had not done much to cultivate research useful to many professions, such as law, medicine, engineering, teaching, planning, administration, and public policy. The Ford Foundation should thus seek to overcome these deficiencies. At one point the 1949 *Study Report* acknowledged that this technocratic viewpoint could pose a threat to democratic ideals, adding that behavioral scientific knowledge could also help individuals to acquire "an understanding of human behavior, their own as well as that of others." In this way, individuals would be better prepared to "maintain the democratic nature of such planning and control," and to "make adequate personal adjustment to the conflicting and changing demands of modern living."[40] Yet, overall, the report gave much greater weight to the use of behavioral science knowledge by professionals and other elites, while proposing that the foundation should stimulate the development of this knowledge and provide advice to "those responsible for the formulation or execution of policy."[41]

Lastly, Gaither's team emphasized the need for objectivity, among other reasons to help ward off damaging controversy. In light of recent criticisms about the social sciences, the report noted that Ford's support for research on controversial issues could be harmful. It could even raise challenges to the legal status of foundations. And it could thus lead foundations to become "timid." With the aim of shielding the foundation from political attack, the report recommended that the BSP cultivate an "objective approach." The establishment of objective facts also promised to place policy advice about how to realize the goals of a democratic society beyond partisan attack.[42]

Following the 1949 *Study Report*, planning for the BSP commenced but with some hesitancy on the part of the foundation's new president, Paul Hoffman—in late 1950 Hoffman replaced Henry Ford II in this position. Hoffman, who had recently directed the U.S.-led, multibillion-dollar Marshall Plan to rebuild Western Europe, focused on developing Ford's new international programs, but he lacked enthusiasm for the behavioral sciences. At one point he even called this a "good field" if one wanted to "waste millions and get nothing."[43] Looking for someone else to take charge of BSP planning, Hoffman called on H. Rowan Gaither in 1951. Two years later Gaither would replace Hoffman as Ford's president as well. Gaither's singular role at Ford gave special weight to his view that "a natural science like physics" provided an appropriate model for the behavioral sciences.

To help him plan and then run the BSP in line with the collection of commitments discussed above, Gaither selected three scholars with extensive connections to national security agencies. Hans Speier, a German émigré, had worked at the New School for Social Research, where he carried out studies on totalitarianism in modern societies. During World War Two he worked in the State Department, set up and directed the U.S. Foreign Broadcast Information Service, and became a specialist in psychological warfare and mass communications. After the war, Speier joined the Panel on Human Relations and Morale of the military RDB's Committee on Human Resources. From 1948 to 1960 Speier also directed RAND's Social Science Division, where he became well acquainted with Gaither.[44]

Gaither's second behavioral science consultant, Donald Marquis, served as chairman of Michigan's Psychology Department and had worked on Gaither's study team in the late 1940s. During World War Two Marquis directed the National Research Council's Office of Psychological Personnel and served as executive secretary of the army-navy Office of Scientific Research and Development's Vision Committee. Marquis also contributed to the massive wartime study of soldier attitudes led by Samuel Stouffer. High-level defense assignments continued into the postwar years, as Marquis became the first chairman of RDB's Committee on Human Resources. By midcentury Marquis stood at the top of his profession, as signaled by his election as APA president.[45]

Following the recommendation of Gaither, Speier, and Marquis, Ford then hired Bernard Berelson to develop specific plans for implementation, and Berelson stayed on as BSP director when the program began full operations in early 1952.[46] For his dissertation at the University of Chicago, carried out under the supervision of psychological warfare expert and library scientist Douglas Waples, Berelson had employed a mixture of quantitative and qualitative analysis to analyze messages from the 1940 presidential campaign, with the larger aim of illuminating the function of mass communication in shaping public opinion and with an emphasis on the positive ways government could use mass communication for this purpose. During World War Two, he worked in the Foreign Broadcast Information Service under Speier. During the early Cold War years, his quantitative social research on political elites and voting behavior brought him national recognition, as he joined the likes of Speier, Leo Lowenthal, Paul Lazarsfeld, and Wilbur Schramm in the booming fields of communications studies and public opinion research. Altogether Berelson would write and edit a dozen books and some ninety articles. Before going to the BSP, his professional posts included project director at Columbia University's Bureau of Applied Social Research (1944–1946), dean of the University of Chicago's Graduate School of Library Science (1947–1951), and president of the American Association of Public Opinion Research (1951–1952).[47]

Though Berelson acknowledged some variations in how different sciences developed, he accepted the notion, common at the time, that all sciences follow

FIGURE 7. Bernard Berelson. *Photo © 2011 The Population Council, Inc.*

a general developmental pattern. Progress in the social sciences, Berelson wrote, began with "a period of 'grand' theoretical speculation," as found in the works of early pioneers such as Adam Smith, Karl Marx, Auguste Comte, and Sigmund Freud. Later, "concentration upon the collection of empirical data" revealed "weaknesses in the early speculative frameworks." Eventually the social sciences, like all sciences, would reach "a more mature state of theoretical development based upon established propositions . . . (supported by) empirical verification." In this last stage, researchers would seek to establish "generalizations about human behavior that are supported by empirical evidence collected in an impersonal and objective way," that are "capable of verification by other interested scholars," and that are dependent on procedures "open to review and replication."[48] As the next chapter shows, during these same years NSF's first social science policy architect, Harry Alpert, presented a similar viewpoint about scientific progress.

To sum up, an extensive planning process that started with the 1949 *Study Report* and continued with input from Gaither, Marquis, Speier, and then Berelson established a fairly coherent vision for pushing the Ford Foundation's fledgling behavioral sciences project forward.[49] This vision posited a universal scientific methodology and epistemology, embraced a natural-science model of inquiry, favored interdisciplinary studies, and underscored the need to develop the field's basic resources. Planning discussions emphasized the need to concentrate on basic research grounded in an objective approach, though with an eye toward practical applications carried out by experts in the professions and in government. The large private foundations of the interwar era and especially the LSRM had tried a similar approach but with mixed results. Foundation programs and policies had evolved through a series of twists and turns, as the commitment to building up academic resources clashed with the desire to concentrate on pressing practical problems, such as social dislocation, depression, and war, and as doubts about the wisdom of pursuing objective, value-neutral social inquiry intensified in the context of an alleged worldwide moral crisis and the frightening Cold War conflict. Against this background, perhaps one should not be surprised to learn that the new BSP would face difficult problems from the outset.

Internal Challenges and McCarthyite Attacks

Hoffman's denigrating comment about wasting millions on the behavioral sciences reflected a wider skepticism about BSP's scholarly orientation among Ford leaders. Some trustees and personnel had little enthusiasm for the field. Though the reasons for their skeptical views differed, taken together their concerns underscored the persistence of troubling questions about intellectual worth, practical value, and political meaning. In addition to these concerns within the foundation, the commitment to social inquiry with an objective and nonpartisan orientation failed to protect the foundation from congressional attacks during the McCarthy Era. Of special interest, a somewhat outlandish though still scathing investigation led by Tennessee representative B. Carroll Reece would present a collection of common charges from conservative politicians and scholars that linked the Ford Foundation's involvement with the social sciences to the advance of scientism, social engineering, moral relativism, socialism, and a host of other purportedly subversive issues.

Besides Hoffman, some foundation leaders had definite reservations about the social or behavioral sciences. These included the physicist Karl Compton, the only scientist on the Ford Foundation's governing board and, as noted already, a member of the natural science elite who opposed the social scientists' bid for inclusion during the postwar NSF debate. From a theologically informed, humanistic standpoint, Robert Hutchins, whom the foundation recruited to direct its educational activities, challenged the social scientist's tendency to turn away from big issues and normative analysis in favor of strictly empirical studies wedded to a disinterested scientific stance. Frank Sutton, a Harvard-trained sociologist and BSP staff member, later recalled that Hutchins thought the BSP was "useless," "utter nonsense."[50]

But trustee Donald David's concern about BSP's abstract character proved to be especially obstructive. Beginning in 1953, David served as the vice-chairman of the foundation's board, the chairman of its executive committee, and a board member at the Ford Motor Company. David had so much influence that Bernard Berelson often asked for his opinion on specific BSP proposals before bringing them before the board. Proposals that David opposed had little chance of approval.[51] Yet David did not object to social science research per se. As the dean of Harvard's Business School, David had made research a central activity. He especially valued problem-oriented studies that promised to facilitate cooperation among the various interests in the modern industrial system, to undermine radical influences within labor, and to provide management with techniques to maintain social harmony and raise industrial productivity. Noting that the world had to choose between "democracy and totalitarianism," David also saw the need for "organized" social research to learn more about "human motives and the forces which govern human relations."[52] But for David, BSP's work seemed "much too abstract," and he charged that Berelson "approached the area of social relations purely from a theoretical point of view." While David recognized Berelson's stature as a great scholar, he concluded Ford had "made a mistake" in hiring him as BSP's director.[53]

Given David's powerful presence and given the Ford Foundation's interest in bringing academic expertise to bear on pressing problems, Berelson's program found itself in a difficult position. When presenting work before the board, other programs that dealt with international affairs, economics, and democracy had an easier time explaining the practical value of their research efforts. In contrast, the BSP, with its dedication to a long-range program for strengthening scholarship, had the "endless problem of defining, explaining, and justifying" its activities, as Berelson put it.[54]

Conservative congressional hostility also promised trouble for Berelson's program. During the Progressive Era, national legislators and the Industrial Relations Commission had attacked the Rockefeller Foundation and other foundations for perpetuating an unjust status quo. But that first wave of antifoundation sentiment ended as the nation entered World War One. Subsequently, the number of foundations soared, from twenty-seven in 1915 to a few thousand by midcentury. This steep growth rate helped revive congressional concern, as did the swelling of anti–New Deal and anticommunist sentiment. Now, congressional conservatives accused the foundations of seeking to destroy the status quo.[55]

In the early 1950s, antifoundation sentiment in American politics and society often went hand in hand with conservative attacks on the social sciences as purveyors of liberal or more radical agendas. Because of the all-too-common confusion between social science and socialism, some patrons and scholars preferred to use the term behavioral science rather than social science: perhaps by leaving out the *s* word (social), they would not raise so many eyebrows. Nevertheless, McCarthyites in Congress and elsewhere regularly accused private foundations and their scholarly experts of supporting un-American activities. As Mike Keen's work shows, J. Edgar Hoover's FBI launched numerous investigations of leading sociologists, including Talcott Parsons, Pitirim Sorokin, and Samuel Stouffer from Harvard alone, while David Price's scholarship reveals the FBI engaged in extensive surveillance of leading anthropologists such as Ashley Montagu and Margaret Mead.[56] Under these circumstances, Berelson feared that "political, economic, and religious opposition" might undermine the nation's tradition of "free inquiry."[57] A series of congressional investigations led by legislators from below the Mason-Dixon Line soon tested the grounds of Berelson's fear.

Georgia representative E. E. Cox, an anti–New Deal leader among conservative southern Democrats, conducted the first investigation. In 1951 Cox introduced a resolution to create a committee to investigate educational and philanthropic foundations and other comparable organizations exempt from federal income taxes. In particular, the Georgia representative focused on foundations involved with "social reform and international relations." He suggested, for example, that the Rockefeller Foundation had promoted communism in China and injected communism into American schools. With strong support from southern Democrats and Republicans, Cox's battle cry to expose the scoundrels "to the pitiless light of publicity" proved popular enough to carry his resolution. The following year the Cox Committee investigation commenced.[58]

Though one would expect the committee's hearings to have produced some high drama, they became remarkable for showcasing a mutual respect between committee members and foundation representatives. This respect emerged partly from foundation leaders' determination to cooperate with the committee and also from the Cox Committee general counsel's firm message indicating that he would not allow the investigation to become a platform for advancing partisan interests. As for the social sciences, conservative philanthropic leaders argued their support did not further leftist ideology but focused, instead, on defeating the communist enemy. The nation's two major Russian studies programs, located at Harvard and Columbia and handsomely supported by the foundations, provided a case in point. As the Carnegie Corporation trustee Russell Leffingwell put it, studying the "diseased" realm of communist thought could make a vital contribution to the war against the Soviet menace: "You must know about it," you must learn "how to vaccinate against it." More generally, Ford president Paul Hoffman claimed that "the best possible cure for radicalism is knowledge."[59]

Hoffman and other foundation representatives further emphasized they would not aid subversive elements in the social sciences—nor in any other area. Under pressure, the three big foundations regularly checked to make sure their grant recipients did not appear on the U.S. attorney general's or HUAC lists of

FIGURE 8. Republican representative B. Carroll Reece with GOP committee members on election night, November 1946. *Left to right*: Clarence J. Brown, B. Carroll Reece, Marion C. Martin, and William C. Murphy, Jr. *Photographer Walter B. Lane/Time & Life Pictures Collection/Getty Images.*

communists or fellow travelers. Raymond Fosdick assured the Cox Committee that the foundations were "in conservative hands."[60]

Satisfied by such assurances, the Cox Committee concluded that foundations and the social sciences represented an important Cold War resource. Agreeing with Edwin Embree and other philanthropic leaders, the committee proposed the foundations should continue to support "controversial" work in the social sciences that could not attract public funding; the Cox Committee *Final Report* also reprinted Embree's article. Thus by providing "risk or venture capital" to advance the "frontiers of knowledge," foundations seemed to be providing an invaluable patriotic service.[61]

The foundations could breathe easier for the time being, but the respite would not last long. The Republican representative and former chairman of the Republican National Committee B. Carroll Reece, who had served on the Cox Committee but attended only one of its many days of hearings, now suggested that the committee had not had sufficient time to study the relevant issues. Reece thus called for another investigation.[62] During the most recent presidential election year (1952), Ford president Paul Hoffman participated in the Eisenhower campaign, while Reece joined a competing and more conservative Republican group that supported Senator Robert Taft and included Reece's close friend and political ally from Ohio, Representative Clarence Brown. During the NSF debate, Brown and Taft pulled no punches in disparaging the social sciences—Brown suggested they were populated by long-haired men and short-haired women interested in telling Americans how to live, while Taft equated social science with politics. Reece's own suspicions about the social sciences would soon become clear. With the Republicans now in control of both legislative branches as well as the White House, a wave of anticommunist investigating frenzy added strength to Reece's cause. As in the case of the Cox Committee, southern Democrats and Republicans provided the votes needed to launch this next investigation.

After acknowledging that the foundations were in the overwhelming majority of instances beyond question, Reece suggested that their size, wealth, and power, and their interests in the social sciences, were disturbing. By mid-century about seven thousand foundations controlled $7.5 billion. This "power of the purse" troubled him, as did the development of an extensive "interlock," whereby many trustees served on the governing boards of two or more foundations, thus allowing the foundations to "act in concert." But Reece singled out their social science activities for special scrutiny. Since he had attended the London School of Economics and then taught economics at a small Alabama college, Reece considered himself quite knowledgeable about this field and the dangers it presented to the American way of life—one suspects the Fabian socialist presence within the London school's social science programs left Reece cold.[63]

According to Reece and his loyal staffers, ever since the New Deal, the nation's intellectual elite had been encouraging a left-wing revolution. Among other resulting problems, this revolution facilitated a great expansion of the federal government and gave the executive branch in particular the major responsibility for maintaining the nation's economic health. Consisting mainly of educators, social scientists, scholarly organizations including the SSRC, and the large private foundations, the intellectual elite manipulated public opinion in order to obtain popular consent for such revolutionary changes. By controlling resources and positions critical in the production and legitimation of knowledge, these revolutionaries promulgated their unorthodox views. Later, they enlisted politicians, college professors, and lawyers in their subversive plot. Consequently, the great American tradition of laissez-faire was losing power as the new welfare state took over, explained the Reece Committee's assistant research director Thomas McNiece. This leftward trend, "wherein the state assumes the ascendancy over the individual and the responsibility for his personal welfare and security," frightened Reece's group.[64]

Reece's inquiry also advanced a pointed critique of social science's value-neutral stance and natural-science envy. While some very influential natural scientists, politicians, and foundation leaders worried that social scientists had not freed their work from value commitments and had failed to follow universally valid principles of scientific inquiry, Reece's investigation pointed in the opposite direction. According to the Reece Committee's research director, Norman Dodd, the pursuit of empirical social research characterized by an emphasis on measuring and quantifying social and individual behavior, along with a corresponding neglect of moral and ethical considerations, had fostered a "deterioration of moral standards."[65] This line of attack drew on long-standing conservative misgivings about the quest for value neutrality in social inquiry, the refusal of many social scientists to address moral issues directly, their tendency to portray religion as part of a premodern, antiquated way of life, and their claim to funding and social influence based on their allegedly neutral, nonpartisan expertise. As noted before, and somewhat ironically in light of the Reece Committee's charges, discussions inside the Rockefeller Foundation and Carnegie Corporation confronted similar issues.

During the early Cold War years, the critique that linked scientism with a host of moral and political ills received further attention from a group of sophisticated writers who would have enormous influence within conservative circles in the decades to come. Their work reinforced what Alice O'Connor has referred to as the "cascading connections between liberalism and godless social science . . . in the ever-expanding but basically consistent litany of charges against foundations and social sciences."[66] In *God and Man at Yale* (1952), William F. Buckley, Jr., argued that modern economics, sociology, and psychology fostered "one of the most extraordinary incongruities of our time," namely, that at Yale University, Buckley's alma matter and an "institution that derives its moral and financial support from Christian individualists," scholars were trying to persuade "the sons of these supporters to be atheistic socialists." The Austrian-born economist Friedrich Hayek also promoted the virtues of individualism, limited government, and free markets. In *The Counter-Revolution of Science: Studies in the Abuse of Reason* (1951), he attacked the "very prejudiced approach" of modern social science, with its "slavish imitation of the method and language of Science," its acceptance of "methodological collectivism" as seen in its reification of such entities as the economy and the state, its corresponding neglect of the individual, and its support for the "conscious control" of social forces, "economic planning," "social engineering," and thus a collectivist society. Taking a longer historical perspective, University of Chicago professor Robert Weaver proposed in *Ideas Have Consequences* (1948) that the "passage from religious or philosophical transcendentalism" explained how "modern man" had "become a moral idiot" and thus helped account for the "dissolution of the West."[67]

As for the Reece Committee inquiry, testimony at the hearings and letters supporting the case against scientism came from a handful of discontented social scientists. These included the Harvard sociologist and Russian émigré Pitirim Sorokin, a colorful, controversial, and prolific scholar who wrote extensively against the ambitions of sociology and related sciences to become "copies of the natural sciences." In the process, Sorokin ridiculed—and angered—many of his sociological colleagues. In a scathing 1941 assessment, Sorokin proposed that in "the Supreme Court of History," contemporary social science would receive "a grade of 'D'" and be placed on "probation." The following decade, in *Fads and Foibles in Modern Sociology and Related Sciences* (1956), Sorokin expanded his attack by exposing what he called the field's "obtuse jargon and sham-scientific slang," "sham operationalism," "testomania," "quantophrenia," "cult of numerology," "grand cult of social physics," "mental mechanics," "wonderland of social atoms," "senescent empiricism," and "sham objectivism."[68]

But it was the sociologist Albert Hoyt Hobbs who emerged as the Reece Committee's favorite social science witness. Hobbs, who had been teaching at the University of Pennsylvania since 1936, except for three years in military service, had recently published *Social Problems and Scientism* (1953). Agreeing with the Reece Committee, Hobbs argued that the great bulk of studies in sociology, history, and political science had severe flaws that stemmed from their misguided outlook. Though the drive to quantify had proven so fruitful in the natural sciences, he said that this drive often led to an impoverished understanding of the individual and social life. In addition, though experimentation had great importance in the physical sciences, experimentation with human beings often violated ethical principles and thus had to be limited. Furthermore, unlike in the natural sciences, social scientists seeking to establish cause-and-effect relationships needed to take into account the issue of free will, though often they did not. Lastly, Hobbs noted that the social researcher could and often did influence the objects of study in ways that the natural scientists could not. Moreover, he found the social impact of such work alarming. When it came to such fundamental matters as "love or sex or patriotism or whatever else the topic may be," Hobbs claimed that scientistic studies produced a "very distorted picture." By drawing a sharp distinction between social investigation and moral inquiry, such work also weakened popular support for traditional values, which led to moral relativity.[69]

Regarding the many damning charges noted above, foundation leaders responded as forcefully as possible, claiming such charges had no basis in reality. As Gaither put it, "the entire program of The Ford Foundation . . . is totally pro-American and actively anti-subversive." Much as other foundation representatives did, he deemed the allegations of un-American activities "erroneous and baseless."[70]

As for the specific charges linking wayward scientism to moral bankruptcy and anti-Americanism, SSRC president Pendleton Herring testified that, in fact, the scientific bent of American social science rested on a firmly established and admirably healthy respect for empirical facts and hypothesis testing. Contrary to the Reece contingent, Herring claimed this scientific mind-set lay at the core of the nation's intellectual and political life, thereby providing an essential foundation for the traditional American way of looking at things. What is more, this healthy mind-set linked America to a proud and deep heritage that could be traced over two millennia, from Aristotle to Francis Bacon to "the father of empiricism" John Locke. Not incidentally, Locke had also "set forth the [political] philosophy most widely accepted by the founding fathers of this country." This deeply rooted "American tendency," continued Herring, sought to discover "the facts . . . to separate mere speculation from factfinding." Indeed, Herring needled Reece, "congressional investigating committees normally follow an empirical approach."[71]

Herring also played what he must have considered his trump card, by asserting that scientific empiricism and objectivity ran directly contrary to the communist way of thinking. In contrast to the great American heritage and tendency, communist thought rested on a blind, dogmatic commitment to a pseudoscientific "Marxian philosophy of history," he declared. "Anyone believing in this predetermined course of affairs or anyone committed to a politically dictated course of policy" could not tolerate "an objective analysis of the facts." Accordingly, Soviet political leaders regarded the objective and empirical approach to social science in the United States "as one of the worst and most dangerous enemies of Communist ideology and Communist expansion."[72] Like an intellectual jujitsu master, Herring tried to redirect the forces arrayed against the social sciences in order to subdue the conservative aggressor. If objective, empirical social analysis posed such a severe threat to the communists, perhaps critics such as Reece were the real un-Americans!

Despite Herring's spirited defense, the Reece Committee's final report judged the foundations and the social sciences guilty of promoting a variety of insidious trends and doctrines, including scientism, moral relativism, social engineering, socialism, collectivism, internationalism, and world government. Representative Reece singled out Herring's defense of empirical, objective studies for special attack, claiming such an approach ignored the central role of "certain basic moral and jurisprudential principles," such as those found in the U.S. Declaration of Independence and Constitution.[73] Adding insult to injury, the report included a two-hundred-page appendix filled with information and allegations concerning dozens of social scientists of strong liberal persuasion, including Ruth Benedict, W.E.B. Dubois, Otto Klineberg, Robert Lynd, and Robert MacIver, as well as charges against the SSRC and many other scholarly organizations cited by the U.S. attorney general and various anticommunist investigating committees. A few years later the Reece Committee legal counsel,

Rene Wormser, presented a similar litany of accusations in his book *Foundations, Their Power and Influence* (1958).[74]

Back at the Ford Foundation, this episode caused serious headaches. On the one hand, careful analysis of public opinion by Gaither's special assistant Waldemar Nielsen revealed substantial discontent with the Reece Committee and its resort to McCarthyite tactics. In addition, a defiant minority report from the Reece Committee's two Democrats, Gracie Pfost (Idaho) and Wayne Hays (Ohio), called the entire investigation "a complete waste of public money." Nielsen himself reasoned that the foundations had run the "gauntlet" as well as possible.[75] According to S. M. Amadae, Gaither also "successfully fought charges of communist leanings and of the foundation's infiltration by communists" during these years. This assessment rings true to a certain extent, because Ford did continue many of its activities and its leaders, including Gaither, maintained their good standing with many elite decision makers in the business, political, and military communities.[76] Nevertheless, as Nielsen reminded his organizational superiors at the time, "a gauntlet is a gauntlet, not a testimonial dinner." Moreover, he also pointed out that the foundations had not made much progress in "neutralizing—or even facing up to—the fundamental suspicions which seem to constitute U.S. public opinion," including the point that philanthropy no longer focused much on popular areas like health and medicine but now provided extensive support for the social sciences. For many Americans, such support meant that the foundations were promoting "socialism, atheism or both."[77]

Thus, criticism of the social or behavioral sciences inside and outside the Ford Foundation created many problems for the BSP. The program's founding vision emphasized the need to develop basic resources in the behavioral sciences rooted in an objective and scientistic approach. However, a number of foundation leaders including trustee Donald David questioned BSP's value. McCarthyite attacks made matters more difficult. Moreover, such attacks looked like they would continue for some time. Near the end of 1954, no fewer than six different congressional committees were considering bills to establish new regulations on foundation activities, and the Reece Committee's final report spoke threateningly of "the problem of foundation survival."[78] As if these developments did not present enough trouble, yet another congressional investigation, this one led by the aggressive anticommunist Mississippi senator James Eastland, promised to subject the Ford Foundation and Berelson's program in particular to scrutiny. As explored later in this chapter, Eastland's investigation raised additional doubts about the allegedly apolitical, value-neutral orientation of the behavioral sciences and contributed to BSP's faltering reputation within the foundation.

Making the Peaks Higher

Meanwhile, the behavioral sciences also encountered various detractors within academia. These included humanistic scholars who challenged the scientific pretensions of the social sciences, as indicated, for example, by this pointed observation from a Ford-funded 1954 report on the behavioral sciences at the University of Chicago: "It would be our mistake, and a loss to all, if we let the phrase 'behavioral sciences' impose on us an exclusive choice of the scientific model, and if that model were to exclude as additions to our understanding the historical and humanistic studies."[79] Equally important, opposition arose from economists who remained coolly aloof from the upstart behavioral sciences. Recent scholarship has documented a widespread trend within postwar economics to resist interdisciplinary efforts, except when economists could be in the driver's seat.[80] Regarding the Ford Foundation in particular, Jefferson Pooley and I have shown that discussions about the possibility of extending the behavioral science perspective to include economics or, even more ambitiously, to reform economic theory and research did not gain much support from economists. Instead of aligning themselves with the new behavioral science movement, economists focused on influencing and obtaining support from Program III, which, as they hoped, would concentrate primarily on funding economics. Program IV also supported some work in economics as part of its broader interest in third world development. The important point is that economists, with few exceptions, preferred to work with their own kind and to pursue research within a disciplinary framework committed to a rational-actor model, rather than to engage in interdisciplinary efforts with, say, sociologists or psychologists, many of whom would not have been inclined to privilege the economist's perspective and saw little ground for its rationality assumption.[81]

Knowing about these clashes within the scholarly world, one might find it remarkable to hear that Berelson's program managed to accomplish much at all. Nevertheless, amidst the McCarthyite political attacks, persistent criticisms within the foundation, and challenges from less-than-enthusiastic sectors of the academic community, the BSP became a major influence in promoting the behavioral sciences as a vibrant field of scholarly activity. All together, the program made a total of 373 grants for nearly $43 million. Not including one huge $13 million grant to create a separate mental health program, the BSP awarded 218 grants to eighty-five institutions for nearly $30 million. In Berelson's words, these grants focused on building up "the basic resources of the behavioral sciences—in ideas, in methods, in men, and in institutions."[82]

Following a well-known strategy in foundation circles, the BSP aimed to advance the behavioral sciences by making the peaks higher. "If you want to turn the system around," explained Berelson, "then you better try to turn it around at the top," and in the behavioral sciences, the top "covers a very large proportion of the field."[83] The BSP approved less than 10 percent of the applications received, allowing Berelson to be highly selective in his effort to make the peaks higher.[84]

Berelson's circle of advisers helped ensure the BSP's significant, albeit contested, influence within the upper echelons in academia. Besides Donald Marquis and Hans Speier, Berelson's advisers included the psychologists Jerome Bruner, George Katona, Lyle Lanier, George Miller, Henry Riecken, and Ralph Tyler; the sociologists Paul Lazarsfeld, Robert Merton, Talcott Parsons, Edward Shils, Samuel Stouffer. and Francis Sutton; the political scientists Arthur Goldhamer, Harold Lasswell, Nathan Leites, Ithiel de Sola Pool, and David Truman; and the economists Kenneth Arrow, Kenneth Boulding, Oskar Morgenstern, Fritz Roethlesberger, Walt W. Rostow, Herbert Simon, Joseph Spengler, and W. Allen Wallis. (Compared with many of their disciplinary colleagues, this group of economists, especially Boulding, Rostow, and Simon, held favorable attitudes about working with other social science disciplines.) Besides their status as leading scholars, many of these advisers worked at elite universities, such as Harvard, Chicago, Yale, California-Berkeley, Stanford, and Columbia. Most and perhaps all of them had close ties to national security agencies as well.[85]

To push the behavioral science project forward, the BSP focused on various levels of the academic landscape, starting with institutional development. Before completing plans to commence BSP operations, the Ford Foundation had already granted about $3 million, divided among thirteen top universities and the SSRC, for the purpose of developing behavioral sciences personnel and institutional resources. Subsequently, the BSP supported a group of self-studies with grants of $50,000 to six institutions: Chicago, Harvard, Michigan, North Carolina, Stanford, and Pennsylvania. According to an internal evaluation, these six comprised a "representative cross-section" of the nation's best university settings for the behavioral sciences. Based on the self-study results, the BSP then made additional grants to these same institutions totaling over $1.5 million.[86] Another major grant provided $750,000 to create a Social Relations Department at Johns Hopkins.[87] As discussed in the next section, the BSP also awarded substantial sums to a number of organizations and research centers that had a significant applied or policy orientation.

BSP's most important and lasting achievement was, arguably, the Center for Advanced Study in the Behavioral Sciences, at Stanford University. Even Donald David recognized the center as the one good thing resulting from BSP's efforts.[88] To establish the center, Berelson's program awarded a 1954 grant for $3 million, slightly more than half of BSP funds allocated that year. From then on, the center remained the single largest item in BSP's budget, receiving a total of $10 million.

The center's planning began with a proposal prepared by a group of high-profile scholarly consultants led by Ralph Tyler. A Chicago-trained specialist in educational experimentation and achievement test construction, Tyler had directed the armed forces examinations staff during World War Two, then became dean of Chicago's Social Science Division and a member of SSRC's board of directors (1945–1949). His star-studded planning group, which included Lazarsfeld, Merton, Marquis, and Stouffer, drew on a Carnegie Corporation/SSRC–sponsored study by Elbridge Sibley that concluded new social science recruits, contrary to popular belief, were "not greatly inferior in previous scholastic achievement, and presumably in innate ability, to most of those in natural sciences." Nevertheless, Sibley argued the supply of adequately trained social scientists did not meet the needs of postwar America. Furthermore, the inadequate supply reflected limitations in university training programs.[89] Tyler's group discussed the need for more adequate advanced training opportunities, but under Tyler's leadership the center would focus on providing scholars at various career stages with the freedom to pursue their work in a stimulating environment.

According to the planning group, the relationship between the center and the nation's universities would be critical in advancing BSP's goals. If all went well, the center's fellows, after one year free from teaching and committee assignments, would return to their universities and create a "multiplier effect," thereby enhancing the center's impact on American higher education. Earlier in the century, the private foundations embraced this same strategy in medicine and other fields.[90] Thus, following the making-the-peaks-higher strategy, Tyler told Berelson that by "helping the top-level talent" to become "still more competent," the center would "tend to affect first-class institutions in large measure." In turn, some graduates of these institutions would become influential faculty members in "other institutions throughout the country."[91]

With Tyler serving as its first director, the Center for Advanced Study in the Behavioral Sciences opened in 1954, supporting thirty-six resident fellows that first year and about fifty in subsequent years. Tagged by one Stanford president as "the think tank up [on] the hill," and referred to by one sociologist as a contribution to "the leisure of the theory class," the center offered an ideal site for year-round study.[92] Its buildings, only one story in height, were constructed of redwood and glass in a modern style typical of that region and time period. From the center's grounds, which sat on a knoll one mile from the Stanford campus, one could peer across San Francisco Bay to admire the coastal mountain range that includes Mount Hamilton.[93]

Given the center's origins, its rigorous process of fellowship selection, and its heavy dependence on Ford funds throughout the 1950s, the work pursued by its resident scholars naturally tended to fit well with BSP's general outlook. According to Ford Foundation records, the fellows represented the cream of the field, "the ablest people in the behavioral sciences."[94] Mathematical work received special attention, and the center became home to the Institute for Mathematical Studies in the Behavioral Sciences. The center also became an important site in the early development of game theory and its spread into academic disciplines beyond economics. In their 1957 book *Games and Decisions*, authors R. Duncan Luce and Howard Raiffa noted they had spent "a delightful year" at the center, during which they pursued their ideas and benefited from the center's clerical support. In addition, the center helped bring together an interdisciplinary group of scholars, led by the brilliant polymath Ludwig von Bertalanffy, who developed general systems theory by combining mathematical approaches and organismic thinking. The spread of behavioral science perspectives and quantitative methods into political science, demography, social psychology, and communications studies also received strong impetus from the center and its fellows.[95]

Complementing the large institutional grants noted above, the BSP helped to advance individual research programs. Berelson said he had no long-range or detailed plan here, only a commitment to "the programmatic support of good men"—again aiming to make the peaks higher. Despite the much larger size of the institutional grants, Berelson concluded that the hundred or so small grants-in-aid for "fundamental research" to individual scholars turned out to be an especially effective use of funds. Most grants amounted to no more than $5,000, though some distinguished scholars including Lazarsfeld received multiple grants and greater sums.[96]

In addition, the BSP sometimes worked with other Ford programs and the SSRC to promote the behavioral sciences. Developing a solid relationship with the SSRC made good sense for reasons indicated earlier: because the council's members represented all the established social science disciplines, because the council had a long-standing interest in advancing the scientific foundations of social inquiry, and because Pendleton Herring, a political scientist and strong ally of the behavioral sciences, served as the council's president. During the post–World War Two years, the council also worked hard to establish the scientific legitimacy of the social sciences in the eyes of natural scientists, government officials, and the public, as discussed in chapter 1. By the mid-1950s, the Ford Foundation, which had overtaken the Rockefeller Foundation as SSRC's major sponsor, was contributing about half of the council's yearly funding base. Between 1956 and 1969, Ford gave the SSRC $23.5 million, almost 75 percent of the council's total outside revenue.[97]

Besides covering SSRC operating expenses, Ford funds supported a number of council committees, undergraduate and graduate fellowship programs, and summer seminars. In these cases, strengthening the quantitative foundations of social research had high priority once again. For example, the Ford-supported SSRC Committee on Mathematical Training of Social Scientists recommended that American undergraduate social science students study "logic and set theory, relations, axiom systems and mathematical models, functions, calculus, probability, and matrix theory." After that committee expired in 1958, the SSRC created a more general Committee on Mathematics in Social Science Research, which gave rise in 1964 to a more permanent organization, the Institute of Mathematical Social Science.[98]

The preceding discussion shows how Berelson's program helped to build up the basic resources in the behavioral sciences at various levels. Summing up the great changes underway, Berelson wrote that scholars at an earlier stage had employed "historical, theoretical, and philosophical terms and wrote treatises," but now they carried out "research projects on specific subjects" and reported their "findings." The overall trend comprised nothing less than a transformation from "scholarship" to "science."[99] To develop more fully an appreciation of what the program's scientistic orientation meant for social science scholarship, the remainder of this section focuses on the foundation's crucial role in stimulating the "behavioralist" movement in political science. Behavioralists became well known for advocating a brand of scholarship that claimed the scientific high ground, which critics of scientism found troubling. Yet certain strands of behavioralist scholarship also supported the foundation's interest in research that promised to advance the national welfare and America's special role in the world, not merely scholarly understanding.[100]

Though this movement had roots in the pre–World War Two era, behavioralists did not become a major force within political science until the 1950s, owing in no small part to the intellectual encouragement, organizational leadership, and financial resources provided by the Ford Foundation in coordination with the SSRC. BSP director Bernard Berelson and SSRC president Pendleton Herring each made important scholarly contributions in the behavioralist spirit. Herring and Berelson also supported the role of their organizations in advancing behavioralism in political science. Ford-funded SSRC Committees on Political Behavior and Comparative Politics, led by the political scientists V. O. Key and Gabriel Almond, respectively, provided behavioralism with crucial support during the 1950s and 1960s. In addition, during this period of time six members of SSRC's Committee on Political Behavior and a handful of other behavioralists served as president of the American Political Science Association.[101]

Sharing much in common with BSP's general standpoint, behavioralist leaders advocated an approach to political studies that would place their discipline firmly within the scientific rather than the humanistic sphere. As Herring put it as early as 1945, the profession of political science had been "essentially a product of the humanistic and philosophic tradition of the liberal arts college." But the time had come for a radical shift, to make the discipline truly scientific. With this goal in mind, Herring proposed that political science needed "an equivalent of the natural scientist's laboratory." According to Key, moving the discipline in this general direction meant that political scientists should concentrate on formulating "modest general propositions" and testing empirical hypotheses with the help of systematic data collection and, wherever possible, quantitative methods of analysis. Emphasizing the need to study recurring political processes through systemic analysis, behavioralists also sought to move beyond particularistic historical, institutional, and legal studies. Instead of analyzing this particular political institution or that unique historical process, many of these scholars found inspiration in the midcentury enthusiasm for social systems thinking, manifest in their studies of political systems and their quest to discern the laws, or at least general principles, that governed their behavior. Consistent with the wider behavioral science project, behavioralists also embraced the drive to eliminate critical theory, normative inquiry, and political ideology from the social sciences.[102]

Yet behavioralists did not stand apart from the great issues of the day, for important strands of their work promised to advance the cause of liberal democracy within the United States and throughout the world. We have already encountered a colorful rendering of this position during the Reece Committee hearings, when Herring defended empirical social inquiry as an excellent way of getting at the facts, as deeply rooted in the American political and intellectual traditions, and as directly opposed to Marxist dogma and communist ideology. Many other behavioralists also asserted that the American system rested on a healthy, tolerant, and pragmatic approach to politics, in contrast to the fanaticism, tyranny, and brutality associated with Nazism and communism. These scholars did not propose Americans had withdrawn from political struggles altogether. But they did suggest that such struggles typically took place within a limited arena of debate, thus undermining the threat of systemic instability and dramatic upheavals. American democracy corrected its problems by "muddling through," based on a process of "incremental adjustments," claimed Yale's Charles E. Lindblom. "The Anglo-American political systems are characterized by a homogenous, secular political culture . . . a multi-valued political culture, a rational-calculating, bargaining, and experimental political culture," explained Stanford's Gabriel Almond. As a result, American politics had an admirably evolutionary character, not a revolutionary one.[103] Meanwhile, a number of European émigré scholars together with homegrown conservative thinkers and legislators continued to worry that the scientistic approach facilitated the spread of moral relativism, undermined support for traditional liberal values, and thus left the nation susceptible to totalitarian ideologues.

To sum up, the BSP did succeed in strengthening the scholarly resources in the behavioral sciences along the lines specified in the 1949 *Study Report* and other early planning documents. Through BSP's provision of substantial resources, its reliance on prominent scholarly advisers, its selective support of institutions and individual scholars, and its efforts involving other Ford programs and the SSRC, the BSP became a leading patron of the behavioral sciences. The sociologist Robert Merton's claim from the early 1960s that "this is An Age of Behavioral Science just as it is, of course, an Age of the Atom and an Age of Automation" contains some truth. So does the sociologist David Sills's assertion from around the same time that Bernard Berelson probably "more than any other person" was "responsible—in the 1950s—for the institutionalization of both the name and the concept of the behavioral sciences in the United States."[104] Ford's promotion of the behavioral sciences helped marginalize alternative investigative approaches as well. Thus it is no coincidence that as behavioralism became dominant in political science, "normative and historical political theory, constitutional law, and political history . . . witnessed a decline in interest," as Peter Seybold has noted.[105]

However, as we have also seen, the example of behavioralism begins to complicate the story of the Ford Foundation's involvement with the behavioral sciences. Important lines of behavioralist scholarship found that, despite some imperfections, American society rested on admirably liberal and democratic foundations and said that the American system of government gave its citizens a reasonable degree of hope that their interests would be taken into consideration. Behavioralists could and indeed did claim that these and related conclusions rested on theoretically informed empirical analysis and thus insisted their work should not be confused with the older tradition of normative political philosophy. Nevertheless, they tackled what they understood to be the big questions of the day about the merits and drawbacks of American democracy and, as discussed in the next section, about whether the American system offered a useful model for developing countries around the world. As the next section also explains, while the Ford Foundation's support for such practically oriented scholarly research remained strong, its commitment to building up the basic resources in the behavioral sciences waned rather quickly. Soon after the Reece Committee's scathing attacks and in response to mounting criticism within the foundation, Berelson's program took a greater interest in applied work.

Blurring the Boundary between Research and Action

Roger Geiger has claimed that Berelson's program remained "fairly secure" as long as Gaither served as Ford Foundation president, until 1956.[106] Gaither certainly contributed to the articulation of the behavioral science perspective, to BSP's early development, and then to the program's protection during a number of difficult years. However, Gaither did not have enough influence to prevent McCarthyite attacks from taking their toll. Nor did he successfully deflect the concerns of skeptics within the foundation. In addition, these factors led to a retreat from the program's early commitment to strengthening the field's scholarly resources. And they encouraged BSP's growing interest in applied research. Geiger's account and the historical literature more generally say little about this noteworthy turn of events, though William Buxton has discussed its relevance with respect to political sociology, a field strongly influenced by behavioralism.[107] Meanwhile, other Ford Foundation programs that dealt with high-profile problems, including juvenile delinquency and third world development, also provided extensive funding for social science with a strong practical or policy orientation. This course of events would help to blur the boundary between social research on the one hand and social action and public policy on the other, especially by the time of the Kennedy and Johnson administrations.

In the mid-1950s, after conducting a study to identify ripe areas of opportunity in applied research, the BSP awarded five grants. The largest provided $731,000 to the Russell Sage Foundation, led at that time by the sociologist, former SSRC president, and 1955 ASS president Donald Young. The 1949 *Study Report* had specifically highlighted the need to develop scientifically sound research that could inform the work of various professions, and Young had a deep personal commitment to this project. Building on this conjunction of interests, the Russell Sage Foundation used its BSP grant to implement reforms in the public health, medical, and social work professions. Recipient institutions included the University of Michigan's School of Social Work, the New York School of Social Work at Columbia University, and Harvard University's School of Public Health.[108]

A number of other BSP grants supported institutions, projects, and individuals engaged in applied work as well. These included one grant for $13 million, by far the largest grant the program ever awarded, to promote the behavioral sciences within the mental health field by expanding opportunities to pursue research careers and by furthering investigations in the areas of psychodynamics, personality development, community mental health, and child disorders. MIT's School of Industrial Management received a BSP grant of $275,000 to support the "application of electronic computer techniques" in the study of "economic systems and industrial organizations." Stanford University accepted a number of BSP grants, including one for $125,000 to create a "clearinghouse" on behavioral science applications directed by communications studies luminary Wilbur Schramm, and another for $400,000 to support three new faculty appointments: a communication specialist, a social psychologist or sociologist, and a mathematician who would contribute to social applications. To develop a new research program on international communication, BSP awarded MIT's Center for International Studies $875,000. And the Population Council in New York got $600,000 to develop research and training in population studies.[109]

Not only did this new applied emphasis represent a departure from BSP's initial orientation; it also made the BSP more similar to other Ford programs that concentrated on policy-related research on domestic and international affairs. In the first case considered here, the BSP began supporting research on the problem of juvenile delinquency, which then became a central focus elsewhere in the foundation. During the 1950s national concern about this problem spread from coast to coast. Troubled youths figured prominently in the classic Hollywood James Dean movie *Rebel Without a Cause* and in Leonard Bernstein's Broadway musical hit *West Side Story*, later turned into a Hollywood movie. Worries about juvenile delinquency mounted within local, state, and national law enforcement agencies and also attracted significant congressional attention. The troubles of American youths became a hot topic of social science research as well. Early in the decade, Berelson's program provided over $350,000 to Harvard University scholars Sheldon and Eleanor Glueck for their work on diagnosing individual susceptibility to this malady. Later in the 1950s, the subject attracted interest at the Ford Foundation's new Public Affairs Program (PAP). Between 1957 and 1961, Ford funding for a combination of scholarly research on juvenile delinquency and related social action programs increased dramatically, from $304,000 to $11.2 million.[110] Moreover, youth development became a major focus within PAP's expanding efforts to tackle a wide range of inner-city problems, including urban poverty.[111]

Directed by Harvard-trained political scientist and liberal urban reformer Paul N. Ylvisaker, the PAP adopted an environmental perspective on juvenile delinquency, in contrast to the work of the Gluecks. Of special importance, the PAP funded investigations carried out by two sociologists from the New York School of Social Work, Lloyd Ohlin, who had a Ph.D. from the University of Chicago, and Richard Cloward, who completed his doctoral studies under Robert Merton at Columbia. Focusing on the structural conditions that shaped the lives of lower-class urban male adolescents, Ohlin and Cloward proposed that delinquency resulted from a lack of sufficient opportunities for youths to join the dominant middle-class culture. Accordingly, they suggested that "the target for preventive action" should not be "the individual or group that exhibits the delinquent pattern" but "the social setting that gives rise to delinquency." In order to "eliminate delinquency," one needed to focus on the social setting that produced delinquency and, specifically, on "the reorganization of slum communities." Toward the decade's end, this work inspired ambitious experiments in social change carried out by Mobilization for Youth (MFY), a new community-based organization in New York City. MFY also acquired substantial Ford funding, while Cloward became one of its codirectors.[112]

The vision supporting MFY saw the social sciences as a solid foundation for social engineering and reform. As explained in detail by the organization's documents, MFY put forth "a broad program of action based on a coherent operating hypothesis and integrated with a carefully designed program of research and evaluation. Essentially, it is a project of social experimentation and investigation, using as its laboratory an urban residential area with a high delinquency rate." Social engineering would, presumably, lead the way toward desirable social change: "the purpose of such a program is to provide evidence of the effectiveness of certain activities directed to altering or changing some undesirable social condition. . . . this proposal should be understood as an effort to bring together the actionist and the researcher in a joint program of social engineering."[113]

The BSP also provided support for scholars and organizations involved in development studies, though the level of support from other Ford programs for policy-oriented work in this area dwarfed BSP's contributions. During World War Two and the early Cold War years, leaders in higher education and government expressed alarm over the lack of expert knowledge about regions outside Western civilization, reasoning that such ignorance posed a major problem for the most powerful country in the world, whose interests now officially extended to the four corners of the earth. The 1947 Truman Doctrine declared U.S. support for "free peoples who are resisting attempted subjugation by armed minorities or by outside pressures." Truman's Point Four Program, announced in his 1949 inaugural address, sought to win the allegiance of third world nations by "making the benefits of our scientific advances and industrial progress available for the improvement and growth of underdeveloped areas."[114] Such concerns informed the 1949 *Study Report*'s call for action on an international scale: with communist influence mounting throughout Asia and Europe, the United States had to devote great resources in order to protect the "free peoples" of the world, especially in "underdeveloped areas," where the people's needs were "vast and seemingly endless."[115]

In the next decade and a half, such concerns informed the Ford Foundation's emergence as the nation's leading private patron of foreign area studies programs. Starting in 1952, Ford took over SSRC's Foreign Studies and Research Fellowship Program, created in 1948 and initially funded by the Carnegie Corporation. This program then became the basis for Ford's enormously influential International Training and Research Program, which had provided over $100 million for area studies in American universities by 1962 and a staggering $250 million by 1967, with about half of those funds going to the social sciences.[116] The BSP contributed some funding, including grants to MIT's Center for International Studies and the Population Council, as mentioned before. As the Cold War unfolded, a number of national agencies joined in, with the State Department, the DOD, the U.S. Information Agency, the Agency for International Development, the CIA, and the Peace Corps all supporting related research and programs, while direct federal support for area studies came with the 1959 National Defense Education Act. Still, Ford continued to provide the majority of funding in the nation's universities. According to a State Department study, 107 of 191 academic centers for research on foreign area studies depended mainly on Ford support as of 1967.[117]

Development researchers from various disciplines including psychology, political science, sociology, anthropology, economics, demography, and communications studies focused on examining the transition from a traditional society to a modern one. Much of this scholarship aimed to identify the key stages of development, critical problems at each stage, and crucial variables that could be manipulated in order to bring about the transition to modernity successfully and, of course, in an America-friendly fashion. One can find many variations on these themes, often reflecting the peculiarities of disciplinary traditions. But it is not unfair to say that, in general, development studies emphasized the establishment of stable anticommunist political systems, the creation of robust economic growth rooted in growing investment and infrastructure, and the development of supportive cultural and educational apparatus consisting of mass media, schools, universities, and other institutions that would encourage the masses to identify with their nation's modernizing goals.[118]

Prominent development scholars also claimed they could help manage American assistance programs effectively, thereby undermining communist control of developing nations and their rapidly growing populations, extensive natural resources, and strategically valuable sites. In *A Proposal: Key to an Effective Foreign Policy* (1957), Max Millikan and W. W. Rostow, two economists from MIT's Ford- and CIA-funded Center for International Studies, reported that the nation's Cold War enemy sought "opportunities to exploit the revolution of rising expectations by picturing communism as the road to social opportunity or economic improvement or individual dignity or national self-respect." The United States thus needed to present "a consistent and persuasive alternative," an attractive model of development that would support American interests.[119] In his 1960 best seller *The Stages of Economic Growth*, Rostow again presented a viable alternative to communist-inspired plans for the third world based on work in development studies and modernization theory. Indeed, the book's subtitle announced Rostow's intention to present his readers with a "Non-Communist Manifesto," though the same basic point appeared in countless works by influential scholars during the 1950s and early 1960s.[120]

During the latter decade, these Ford-funded ventures in the social sciences informed an array of new national initiatives. As the Kennedy and Johnson administrations set out to tackle various domestic ills, Ford staff members and Ford-funded scholars served in various advisory roles. Just to mention a few examples, Paul Ylvisaker chaired a presidential task force on the city, while Lloyd Ohlin served on the President's Committee on Juvenile Delinquency and

Child Crime. During the Johnson administration, the War on Poverty's controversial community action programs reflected the influence of staff members and projects associated with Ylvisaker's Public Affairs Program.[121] In her insightful work on social science and antipoverty programs, the historian Alice O'Connor has rightly noted that Ford had taken on "the role of 'change agent,' demonstration station, and policy incubator." In the process, the foundation's work had brought social science perspectives and associated visions for social change before a powerful new contingent of national leaders "willing to embrace the cause of liberal social reform."[122]

In a similar fashion, development studies contributed to ambitious new foreign policies for the third world. Ever since the Ford Foundation's enormous expansion and transformation in the late 1940s and early 1950s, many trustees and staff members had close connections to the nation's foreign policy elite. Numerous Ford-funded organizations and scholars also had good relations with the nation's military, intelligence, and foreign policy apparatus. MIT's Center for International Studies became an especially influential conduit in the rise to prominence of the nation's so-called action intellectuals, including Rostow, who served as President Kennedy's national security adviser and subsequently as President Johnson's foreign policy assistant. In these high-level government posts, Rostow pressed aggressively for anticommunist modernization programs and counterinsurgency initiatives in the third world.[123] As the historian Michael Latham has argued, these events reveal the "profound role of social science in the exercise of American power and the definition of a national sense of self" during that time period.[124]

Typically, historical accounts of the BSP have paid little attention to its relationship with other Ford programs. But, as we have seen, one needs this wider perspective to understand what happened to BSP's mission of building up the scholarly resources in the behavioral sciences including the promotion of objective, basic research. After only a few years of trying to pursue that mission, the BSP shifted its focus to behavioral science applications. BSP assistant Frank Sutton stated the lesson learned: "never sell a research project as a research project to a board of trustees."[125] Meanwhile, other Ford programs during the 1950s and 1960s provided extensive support for action-oriented scholars who focused on such timely subjects as juvenile delinquency, urban poverty, and third world development. As seen in Cloward and Ohlin's call to reorganize slum communities, in MFY's embrace of experimental social engineering, and in Rostow's and Millikan's proposals for the third world, the work favored by the foundation often combined social research with direct attention to normative concerns about the social, economic, and political orders, both at home and abroad. Such work, in turn, informed the development of community projects, demonstration programs, and public policy initiatives. In short, with the help of the foundation's enormous resources, these scholars offered not merely disinterested academic analysis but visions of social progress informed by social science expertise.

The "Last Straw"

During the mid- to late 1960s, domestic and foreign policy conflicts would enmesh action-oriented scholars in controversies that raised fundamental challenges to the politics–patronage–social science nexus. Heavy criticism came from a growing chorus of disenchanted liberals and more radical voices on the one hand, while conservative critics of social engineering and its alliance with 1960s liberal social reform expressed their concerns loudly as well. Whether from the left or right, critics from inside and outside the ivory tower charged that social science scholarship funded by private and other powerful patrons hardly qualified as value neutral or nonideological. Yet prior to those explosive controversies, Ford's ventures in the behavioral sciences had already attracted political and intellectual scrutiny during the 1950s, especially from conservative voices. Following the difficult Reece Committee inquiry, public uproar over a BSP-funded project inspired yet another congressional investigation. This episode, though largely forgotten today and rarely mentioned in the historical literature, once again challenged the allegedly objective and empirical bent of the behavioral sciences that its promoters so often championed.[126] The troubles began with an ambitious study of the jury system carried out by University of Chicago scholars.

The so-called Wichita Jury Study employed a novel research strategy that involved recording jury deliberations without the knowledge of the jury members themselves, an idea originally advanced by a Chicago Law School graduate and Wichita lawyer named Paul Kitch. The Chicago group of scholars, led by the Law School dean Edward H. Levi (who later became the University of Chicago provost and then U.S. attorney general under President Ford), employed a battery of investigative techniques such as interviewing jurors and conducting experimental mock trials. They focused on doing "field research" rather than traditional "library study." They emphasized their interest in obtaining accurate empirical data in order to test specific hypotheses in a field still dominated by historical precedent and humanistic modes of reasoning. With this end in mind, secret jury recordings would help them to "appraise the realism of the moot, experimental deliberations."[127] As the original 1952 grant application to the BSP put it, the behavioral sciences had an essential role in "a realistic study of the legal system."[128]

This wording invoked the contested history of legal realism and sociological jurisprudence whose proponents, including former Supreme Court justice

Louis Brandeis, had been arguing ever since the- interwar era that traditional legal talk about the application of natural rights and universal principles to specific cases reflected muddy thinking. Legal realists suggested that such an outlook appealed to metaphysical or religious claims and thus could not be subject to empirical examination. In addition, legal studies that focused on the application of general principles to specific cases ignored social realities such as class and race that, in practice, greatly influenced how juries and judges arrived at their decisions.[129]

For the Chicago scholars, everything seemed to be going well at first. They had targeted a field of study loaded with practical relevance and ripe for behavioral science investigation. Two Ford Foundation grants provided a combined sum of $1.4 million. The project's star-studded advisory group included Harold Lasswell, Robert Merton, and BSP's Hans Speier. The researchers worked at one of the nation's elite universities with an outstanding social science tradition. They had proposed a clever new methodological tool. Lastly, after agreeing to protect jurors by deleting any information in the transcripts that could reveal their personal identities, the Chicago scholars had gained a judge's permission to proceed. So, with the use of microphones hidden in the heating apparatus, they recorded jury deliberations in six civil cases. Though they made no further recordings, at one point Kitch's law office mentioned making a "minimum of 500."[130] Democratic realities soon intervened, however.

After a social psychologist from the Chicago group named Fred Strodtbeck discussed the recordings at a conference in Estes Park, Colorado, concern about the study spread rapidly across the country. According to an American Civil Liberties Union representative, the secret recordings violated Article III and Sixth and Seventh Amendment constitutional rights. More specifically, the recordings compromised the principle that a jury should be "as impartial as humanly possible—as free as scrupulous observance of rules can make it from any kind of surveillance, embarrassment, or coercion." In the nation's capital, the U.S. attorney general declared the Justice Department's opposition "to any recording or eavesdropping on the deliberations of the jury."[131]

Conservative outrage over the Supreme Court's 1954 decision in *Brown v. Board of Education of Topeka, Kansas* further darkened the stage. In that landmark decision outlawing segregation in public facilities, the court justices invoked "modern authority" to support the claim that segregation harmed the educational and mental development of black children. Chief Justice Earl Warren's decision cited seven social science studies as the basis of such authority, including Gunnar Myrdal's *An American Dilemma*. According to pro-segregationist critics, the Supreme Court had hastily tossed away decades of sound legal reasoning and undermined the constitutionally grounded power of the separate states—"jurisprudence gone mad"—in favor of dubious social science opinions and a specious scientific doctrine of racial equality, all supported by scholars with liberal biases or worse.[132] Mississippi Democratic senator James Eastland introduced a resolution depicting Myrdal and his collaborators as communist allies and called for an investigation. Though such an investigation never took place, Eastland used his position as chairman of the Senate Internal Security Subcommittee to investigate the Wichita Jury Study.

A good portion of the 1955 subcommittee hearings focused on determining whether or not the Chicago researchers were communists or fellow travelers. The Mississippi senator hounded Chicago's Harry Kalven, director of one part of the jury study, about a letter from Kalven to President Truman requesting clemency for Ethel and Julius Rosenberg, who had been convicted of espionage and then executed. In addition, the chief counsel for Eastland's committee pressed Kalven about an article he wrote defending J. Robert Oppenheimer, who had recently failed to clear his name against the Atomic Energy Commission's charge that he posed a significant national security danger.[133]

Of course Eastland's subcommittee attacked the jury study itself as well. At the hearings, Kalven spoke of "an attack on the jury system for the last hundred years in America" and mentioned "a memorandum" explaining the nature of this attack. Eastland saw the situation differently, asserting that secret recordings represented a threat to the jury system, which, in turn, posed a danger to the nation's internal security. Did Kalven not believe the "American jury system" represented "one of the greatest safeguards of human liberty?" Did Kalven not recognize that the "American jury system is superior to the Soviet court system?" Did Kalven agree with Karl Marx's associate Friedrich Engels who once called the English jury system the "culmination of juridical falsehood and immorality?" More generally, Eastland wanted to know if the Chicago scholars meant to subvert American constitutional law by destroying one of its cornerstones.[134] A few years later Rene Wormser, formerly the Reece Committee's legal counsel, called Kalven and Levi "fictitious liberals," a derogatory phrase associated with the anticommunist tactics of J. Edgar Hoover.[135]

Though the subcommittee produced little evidence of subversion, Eastland remained unsure that the BSP-supported group at Chicago and like-minded social researchers elsewhere had sufficient respect for the jury system. For good measure, he and his Senate colleague William E. Jenner, an ardent anticommunist Indiana Republican, sought to cut off the malignant growth at its source by sponsoring a bill that would make "recording, listening to, or observing proceedings of grand or petit juries while deliberating or voting" a federal crime. Those found guilty would receive a fine up to $1,000, one year in prison, or both. With President Eisenhower's signature, the bill became law in 1956.[136]

Back at the Ford Foundation, Berelson's program was quickly losing the struggle for dollars and respect. He complained that his superiors went overboard in allowing public relations to shape their decisions. And he reminded them that the foundation should consider big ideas and take risks, something federal patrons presumably could not do as well. The board's negative decisions on a series of BSP-recommended project proposals thus troubled him. One proposal came from left-leaning researchers who wanted to study the sorry plight of American Indians. Another proposal focused on the explosive subject of desegregation.[137]

A third proposal requested support for Alfred Kinsey's pathbreaking research on human sexual behavior. In recent years, conservative critics on the Reece Committee and elsewhere had accused Kinsey of denigrating proper standards of sexual morality and conduct. Amidst mounting public controversy, the Rockefeller Foundation, after backing Kinsey's work for a decade, decided to cut off its support. When Kinsey turned to the Ford Foundation for funding, Berelson responded with enthusiasm. But the trustees did not, probably, reasoned Berelson, because they did not want to assume the public relations risk.[138]

The most troubling case concerned yet a fourth proposal, this one for a new encyclopedia of the social sciences. In the early 1930s, the large foundations sponsored the fifteen-volume *Encyclopedia of the Social Sciences,* a monumental scholarly product of the interwar period. In light of all the new social science work done since then, the encyclopedia's associate editor Alvin Johnson asked the Ford Foundation to support a new one. After studying this proposal carefully, Berelson's team concluded it deserved funding and then developed plans to have a group of top universities initiate, sponsor, and prepare the project.[139] But BSP's recommendation for a $2 million grant met a stone wall, with the powerful Donald David rallying the opposition. David worried that screening all potential participants in this large project adequately would be difficult, even impossible. Of the hundreds of authors who would contribute, a few rotten or at least suspicious apples could provoke public disdain for the entire undertaking. "We had to be sure that we were doing nothing that was political in nature," David later explained. Consequently, despite much hard work, this initiative died as well. Berelson called the board's negative decision "disappointing and regrettable." In a *New York Times* letter to the editor, Johnson complained that the "reckless clay-pigeon shooting of the Reece Committee" had made the Ford Foundation and other foundations "gun-shy."[140]

Retrenchment soon reached its logical conclusion. Gaither departed the Ford Foundation in 1956. His replacement, Henry T. Heald, an engineer and former chancellor of New York University, said he had a "somewhat skeptical" view of the social sciences. Berelson added that Heald possessed a typical "engineering prejudice" against social research.[141] In a 1960 speech to the Brookings Institution, Heald declared that "outside" of the natural sciences, "the path of knowledge has been less sure," often ranking behind "caprice, tradition, or partisan expediency."[142] Without much hesitation, Heald eliminated Berelson's contentious program. In 1957, after making a final series of terminal grants, the BSP met the ax.

A distressed Berelson labeled this decision "unwise" and "hurtful" to the field and the foundation.[143] Although Berelson recognized the importance of public relations, he thought Ford leaders had responded to the charges of subversion too quickly. Turning once again to history for perspective, he reckoned that withdrawing from the behavioral sciences in midcentury America was similar to cutting off support for the physical sciences in the early eighteenth century because they challenged established philosophical beliefs, or avoiding the biological sciences in the early nineteenth century because they threatened religious doctrines.[144]

In sum, McCarthyite attacks, including the little-known Senate investigation of the Wichita Jury Study, underscore the persistence of conservative suspicions of the behavioral sciences and their private patrons during the mid-1950s. Only one year after the Reece Committee's critique of scientism, and in the midst of bitter controversy over the Supreme Court's invocation of social research in the *Brown* decision, Senator Eastland's investigation suggested social scientists had no business conducting a secret study of jury deliberations. Whereas the study's grant proposals and scholars championed its objective and empirical orientation, Eastland's position indicated the jury system was a sacrosanct American institution, whose legitimacy derived from legal-constitutional principles but which the BSP-funded study seemed destined to subvert. In addition, those investigations evidently had a chilling impact, as suggested by Berelson's complaints that Ford trustees had become excessively cautious, by David's recollection of a strong-felt need to avoid projects that might be attacked as political in nature, by the series of BSP-supported project proposals rejected by the trustees, and of course by BSP's closure. Another foundation insider, Richard Magat, later concluded that McCarthy Era investigations had indeed taken "their toll," by encouraging Ford to avoid "anything with great potential controversy" and to concentrate, instead, on safer items.[145] More recently, Alice O'Connor reached a similar conclusion, claiming that Ford and other large foundations distanced themselves from or ended funding "for their more controversial grantees, in effect accommodating their ideological critics in the course of defending their own ideological neutrality."[146] As the previous section indicated, it would be a mistake to conclude more generally that the foundation backed away from social research. An array of subjects with strong policy relevance, such as juvenile delinquency and third world development, continued to receive extensive support from practically oriented foundation programs. But political pressures in combination with internal disfavor did curtail the foundation's commitment to strengthen scholarly resources in the behavioral sciences per se, as seen through the BSP's rocky and short-lived existence.

Conclusion

The modern Ford Foundation and the BSP in particular did much to create a national environment favorable to the takeoff of the behavioral sciences. Despite the twists and turns in foundation support for the social sciences during the interwar era, and notwithstanding the serious doubts raised about scientism and the alleged worldwide moral crisis during postwar discussions at the Carnegie and Rockefeller foundations, the 1949 *Study Report*, BSP planning documents, and the BSP itself set forth a vision that assumed the social sciences could and should take the natural sciences as their model. A corresponding social engineering outlook supposed basic and objective studies would produce knowledge that professional experts and national leaders needed in order to address serious domestic problems and worrisome foreign policy challenges. To push the behavioral sciences forward the BSP followed a strategy common in philanthropic circles; namely, Berelson's program focused on the top of the academic hierarchy, based on the assumption that making the peaks higher was the best way of influencing the entire academic system. The BSP thus channeled its resources in a highly selective manner, providing valuable support to the Center for Advanced Study in the Behavioral Sciences, to major social science organizations including the SSRC, to programs and research centers at elite universities, and to dozens of topflight scholars.

That the foundation's outlook on the behavioral sciences had much in common with that of the national defense establishment is no coincidence. Berelson, his main assistants, including Marquis and Speier, and his wider circle of prominent academic advisers had close ties to defense science programs, as did H. Rowan Gaither. In addition, much like prominent social scientists who sought inclusion in the proposed NSF or in the defense science establishment, behavioral science advocates found themselves under pressure to explain their scientific basis, practical value, and political acceptability. Given doubts among the foundation's leadership together with the prominence of conservative critics in the political and scientific arenas, BSP's architects believed that the pursuit of basic studies with a presumably objective, apolitical character would have strategic value because it would protect the foundation from charges of partisanship and, even worse, subversion.

Nevertheless, as we have seen, a series of investigations during the McCarthy Era advanced the charge of subversion. Though the Cox Committee inquiry started out with some damning allegations about the role of private foundations and the scholarship they supported, the investigation produced a surprisingly favorable assessment and even encouraged the foundations to supply the social sciences with additional risk capital. However, the Reece Committee accused the large foundations and a bevy of foundation-supported social science organizations, projects, and scholars of adopting a scientistic perspective that, in turn, contributed to a host of troubles, including the spread of moral relativism, godlessness, and socialism. Similarly, the investigation of the Wichita Jury Study indicated that this allegedly objective BSP-funded project and its University of Chicago researchers posed a threat to the American jury system and thus had subversive implications.

Moreover, those conservative attacks had a major impact, as Ford Foundation leaders, already skeptical of BSP's scholarly orientation, became increasingly worried about the political repercussions of those attacks. Following the Reece Committee and Wichita Jury Study investigations, the foundation's trustees became more wary than ever of funding projects that might attract scrutiny. At one point, Berelson suggested the latter investigation became the "last straw." Judge Charles Wyzanski, another Ford trustee, thought that the stillborn encyclopedia project was the BSP's death knell.[147] But whatever the precipitating event, the underlying causes of death are clear: a combination of various internal strands of disfavor, a series of McCarthyite congressional investigations, and the threats of further investigations. Consequently, the BSP, the largest and most important single effort by a private foundation to advance the behavioral sciences as a scholarly enterprise in Cold War America, was simply eliminated.

Placing BSP's struggles in the context of other Ford programs highlights additional difficulties that beset the midcentury effort to carve out a space for allegedly objective and value-free academic behavioral science in a way that would distinguish it from social action, partisan politics, and policy work. As discussed in the next chapter, the psychologist and influential NSF staff member John Wilson recommended in the early 1950s that social scientists should not engage in "action research," as he believed such research led scholars to confuse the distinct roles of social researcher and social reformer. Whether or not one agrees with Wilson's position, a great deal of Ford-funded social research fit this description. Even the BSP itself underwent a reorientation toward applied work, after its efforts to fund more academically oriented projects became increasingly unpopular within the foundation. As seen in the cases of foundation-supported research on juvenile delinquency and development studies, the notion that such research would provide the basis for managing social problems at home and winning the Cold War in the third world helped to blur the boundary between research and action. Even in the case of behavioralism in political science, which Ford and the SSRC did so much to promote, advocates furthered their vision of scientific inquiry as a bulwark against leftist ideology and used their scholarly works to argue for the fundamental goodness of the American political system. Returning to Raymond Fosdick's complaints about the social sciences and the insufficiency of scholarly *analysis*, it seems fair to conclude that following BSP's closure, the foundation continued to fund a great deal of social science research in ways that promised to marry analysis to vision, in order to realize the practically oriented aims of its other programs.

After the BSP's demise, one anonymous scholar noted with dismay that the Ford Foundation was "disavowing any responsibility" for developing "basic knowledge" at the very time that it was demonstrating an increased interest in the "utilization of the behavioral sciences in various fields."[148] However, during those same years the NSF established a primary interest in supporting basic research while disavowing any interest in applied research. Given the fierce debate over social science in NSF's legislative origins, the agency's emergence as an important patron was far from inevitable. At least a spectator in 1950 would have had little reason for believing the new agency would take much if any interest in social science. The next chapter examines the agency's slow, cautious, increasingly important, and yet also persistently troubled engagement with American social science and its scientistic wing in particular.

4

Cultivating Hard-Core Social Research at the NSF

Protective Coloration and Official Negroes

The social sciences—except for a few extremely limited areas—are a source of trouble beyond anything released by Pandora.

—Kevin C. McCann, National Science Board member and Eisenhower biographer, 1958

The social sciences have prospered best in the federal government where they have been included under broad umbrella classifications of the scientific disciplines. . . . in close company with scientific areas which enjoy the prestige and status of biological or physical sciences, the social sciences have enjoyed a protection and nourishment which they normally do not have when they are identified as such and stand exposed, "naked and alone."

—Harry Alpert, sociologist and first NSF social science director, 1960

Although the initial legislative bills from the mid-1940s had anticipated that the new federal science agency would become the comprehensive centerpiece of the postwar federal research system, the lengthy debate over competing proposals gave the military and other patrons the opportunity to establish their own substantial programs. When the NSF began operating in the early 1950s, the agency claimed a comprehensive interest in supporting basic science. But because of its comparatively puny budgets, the agency could hardly fulfill the earlier expectations. Still, the NSF assumed a small but increasingly important role in the nation's commitment to basic or fundamental research, through the support of scientific projects, fellowships, and research facilities. Basic science, which by definition promised utilitarian payoffs only in the long term, if at all, had never been the primary concern of any national agency. In addition, during the early 1950s, the ONR and other defense science agencies that funded some basic research began to place a stronger emphasis on applied science. In this context, the NSF "broke new ground," as its first annual report proudly noted. Vannevar Bush went so far as to declare that the agency's mandate to support basic research qualified as "one of the minor miracles."[1]

This chapter examines the development of NSF programs and policies to fund the social sciences up through the early 1960s. During this period of time, the agency's support for these sciences grew, at first slowly and cautiously, then in later years with increasing vigor, but always within well-defined boundaries that emphasized the strictly scientific character of NSF-funded research. As we will see, the agency's approach had much in common with the position advocated by SSRC scholars during the postwar NSF debate. But the immediate origins of its approach lay in the continuing struggle of social scientists to establish their scientific credentials within the federal science system in general and within this natural science–oriented agency in particular. The young agency's leaders, including its first and longtime director Alan Waterman, a physicist, approached social science funding with caution, much as Vannevar Bush had. In addition, anticommunist politics in the nation's capital during the agency's early years placed additional pressure on the NSF to distinguish agency-funded research from forms of social inquiry associated with suspect ideologies and left-wing politics.

Under these conditions, the sociologist Harry Alpert, who worked at the NSF from 1953 to 1958, became the agency's first social science policy architect. A Durkheim scholar, Alpert had a deep interest in the history of the social sciences, in the social contributions of these sciences, and in the relationships between the social and natural sciences. In various writings, Alpert, who has received little attention in the historical literature, expressed support for a rather broad range of scholarship in the social sciences, including work that sought to assess and reform society in order to achieve liberal democratic ideals. However, in response to political and institutional pressures that made social science funding problematic at the agency, Alpert crafted a carefully circumscribed strategy that limited NSF support to the so-called "hard-core" end of the social research continuum. For many years to come, agency documents followed his policy guidance by defining such work as research that met universally valid scientific principles, including generalizability, verifiability, and objectivity. By the end of the 1950s, the NSF already ranked fourth among federal agencies based on the amount of funding they provided for basic social research. With the demise of the Ford Foundation's BSP, the NSF also became the nation's only major patron for academic social science that did not take a strong interest in its practical applications. All the while, the agency remained committed to the guidelines and carefully circumscribed approach Alpert had set forth for strategic purposes.

This chapter also explores important criticisms of NSF's social science efforts that once again raised doubts about the scientistic project. Based on a theological perspective, Father Theodore Hesburgh, president of the University of Notre Dame and a member of NSF's governing board, proposed in the late 1950s that the agency should welcome other perspectives within the social sciences that lay beyond their hard-core wing. Though Alpert himself played a major role in establishing NSF's hard-core orientation, after he left the agency he argued that the reigning scientific pecking order harmed the intellectual development of the social sciences and undermined their ability to promote social progress. Meanwhile, from a very different viewpoint some natural scientists and politically conservative board members continued to express serious doubts about the social sciences and the agency's support for them, with Kevin McCann, the author of a glowing biography of Dwight D. Eisenhower, comparing the social sciences unfavorably to Pandora's box. Last, this chapter will examine a controversy over the agency's limited support for research on politics that took place during the early 1960s and was led by the political scientist Evron Kirkpatrick.[2]

Making Space for Potential Troublemakers

The "other sciences," as NSF's 1950 enabling legislation obliquely referred to them, had a rough start in the new agency. Conservative politicians and natural scientists who claimed the social sciences were ideological, political, and thus of doubtful scientific character had already kept them from being directly included in this legislation. Within the Bureau of the Budget, one individual observed in the early 1950s that the NSF was "gingerly approaching" the social sciences, and the agency's leaders seemed "sensitive, overly perhaps," to potential "criticism."[3] But when other federal agencies including the Defense Department asked about NSF's intentions, the agency's leaders decided to hire a social scientist to study the matter. In 1953, Harry Alpert got the call. At the new agency Alpert's role would turn out to have major significance, both because he managed to establish a viable and long-lasting framework that enabled the agency to concentrate on funding hard-core research and because in creating and promoting this framework, Alpert had to bracket some of his own philosophical and political concerns that reflected wider challenges to the scientistic project.

Starting with his doctoral studies and continuing throughout his career, Alpert often presented a catholic view of the social sciences and their social responsibilities. In his Columbia University dissertation, Alpert examined the life and scholarly career of the great French sociologist Emile Durkheim and included an assessment of Durkheim's social philosophy. This study culminated in a 1939 book wherein Alpert examined "Durkheim's broad view of sociology as a positive science of social behavior, as a rejuvenating method of social investigation, and as a solid foundation for an integral social philosophy."[4] Alpert's biography, together with influential studies by Talcott Parsons, helped bring Durkheim's contributions to the attention of a wider American audience.[5] His 1939 book and other writings also established Alpert's deep interest in a range of foundational issues in the social sciences, including the ontology of the social world, social science methodology, relations between the social and natural sciences, and the social relevance of social research.

Though commentaries about Alpert have typically suggested he had few if any qualms about the scientistic framework he helped develop for the NSF, Alpert's scholarly writings reveal that he had an impressively complex understanding of social science that did not mesh well with that framework. As Jefferson Pooley and I have recently noted, Alpert has been "remembered as a quantitative evangelist and advocate for the unity-of-science viewpoint." Yet his writings on Durkheim, on the development of sociology as a profession, on the field of public opinion research, and on federal social research programs reveal that Alpert was, in fact, "an urbane critic of natural-science envy, social scientific certainty, and what he saw as excessive devotion to quantitative methods." Regarding the question of the unity of the sciences, Alpert often recognized that the social and natural sciences had much in common at the levels of epistemology and methodology. But in various places he also pointed out some fundamental differences, for example in his commentaries about the critical value of *interpretation* in social inquiry. Regarding this topic, Alpert praised Max Weber's recognition that social scientists needed to understand the subjective viewpoint of the human subjects they studied. Alpert also noted that this vital need implied significant divides between the social and natural sciences, regarding how they come to know the objects of their inquiries and regarding the types of knowledge they produce.[6]

Alpert further believed social scientists should promote the ideals of human equality and dignity, for example by documenting and developing strategies to combat anti-Semitism.[7] Courses in anthropology with the renowned German Jewish émigré scholar Franz Boas and Boas's student Margaret Mead at Columbia University had contributed to Alpert's advocacy of racial equality and his opposition to anti-Semitism within a broader liberal democratic framework. Geoff Alpert, one of his sons, has pointed out that while his father believed strongly in the value of social science, his wide-ranging scholarly interests included extensive knowledge of literary classics and a commitment to knowledge as the basis not for cultivating amoral technical expertise but for furthering human striving toward the good life and the good society.[8]

After completing his Ph.D., Alpert held a number of governmental and academic posts. During the 1940s, he became involved in the war effort working at the OWI and the Office of Price Administration as a public opinion analyst. After the war, he maintained government ties, through positions as a researcher and coordinator of statistics in the Bureau of the Budget and as a consultant on manpower problems for the air force's Research and Development Board. This government work helped to make statistics and public opinion research central interests for Alpert. He also continued to pursue his career as a university scholar, including posts as an assistant professor of sociology at the City College of New York, a research consultant at Columbia University's Bureau of Applied Social Research, and an associate professor of sociology and chairman of the Department of Anthropology and Sociology at Queens College. At one point Alpert was a leading candidate for an attractive position at the University of Maryland but did not get the job, owing to anti-Semitism, which left him deeply upset. Though Alpert did have a Jewish heritage and had been raised in a Jewish orphanage, his own family did not follow Judaism in a religious sense. At another point, Alpert turned down a handsome job offer from his friend George Gallup. Rather than joining the new and thriving public polling organization founded by Gallup, Alpert preferred to pursue a scholarly career. Before he arrived at the NSF, then, Alpert had considerable knowledge of the political, institutional, and intellectual contexts of American social science.[9]

For Alpert, promoting the social sciences at the NSF would be a compelling but arduous task, largely because the agency focused overwhelmingly on the natural sciences. As of 1953 the agency had three major divisions: (1) Biological and Medical Sciences; (2) Mathematical, Physical, and Engineering Sciences; and (3) Scientific Personnel and Education. Prominent men from the "hard" sciences and top administrators at elite universities dominated NSF's leadership. Many of them also had influential positions within the NAS and defense science programs, including the chemist James Conant, who served as the first chair of the National Science Board (NSB), the agency's twenty-four-member governing body; the bacteriologist E. B. Fred, NSB's first vice-chairman; and the biophysicist Detlev Bronk, the first leader of NSB's nine-member executive committee.[10]

Natural scientists filled many other NSB seats as well, while only one member, Frederick Middlebush, had a social science doctorate. But even Middlebush, a political scientist, had long ceased to be an active scholar, and his most important qualifications for NSB membership were his high-level positions in university administration.[11] A few other NSB members had more than a passing interest in the social sciences, namely, Chester Barnard, Charles Dollard, and Sophie Aberle. But as with Middlebush, their most relevant expertise in this context lay elsewhere. Barnard and Dollard had managed large private foundations, Rockefeller and Carnegie, respectively. Aberle, a physician and one of only

FIGURE 9. Alan Waterman at a reception honoring Vannevar Bush, December 1, 1955. *Left to right*: Waterman, Detlev Bronk, Bush, Chester Barnard. © *Bettmann/CORBIS*.

two women in the original NSB group, had a deep interest in anthropology and especially the culture of Pueblo Native Americans. But at the NSF she primarily represented the medical profession.

The agency's first director, the physicist Alan Waterman, approached the social sciences hesitantly, along the lines of Vannevar Bush's cautious recommendations and reflective of deeper connections between their careers and science policy views. Waterman had worked in OSRD under Bush and also supported Bush's conservative science policy plans in the 1940s. Before moving to the NSF, Waterman had also worked as ONR's chief scientist. Though psychology and some other parts of the social sciences received ONR support, this agency, similar to other defense science agencies, concentrated on the natural sciences first and foremost, as discussed in chapter 2. Moreover, Waterman turned to ONR personnel when recruiting his new NSF staff. Both at the ONR and then during his dozen years as NSF director (1951–1963), Waterman championed the putative separation of science from politics.

The new agency's struggle to acquire political support and financial funding added to Alpert's difficulties. Bush's *SEF* and the original Magnuson and Kilgore legislative proposals had all called for a large and comprehensive science agency. However, the NSF was initially quite small compared with the enormous postwar federal science programs devoted to military, energy, and medical research. As a result, the agency had "pitifully small" budgets, to borrow Alpert's words.[12] The scarcity of dollars made NSF's leaders even less inclined to allocate funds to the problematic social sciences.

On top of these obstacles, NSF leaders feared that including the social sciences might result in efforts by outsiders to control the agency's affairs, an issue that received ample attention in *SEF*, in the drawn-out postwar NSF debate, and in countless discussions and documents during the agency's early years. Extensive national loyalty concerns during the early 1950s gave NSF leaders good reason to worry that heavy-handed intervention could compromise the intellectual freedom of scientists and the academic autonomy of scientific institutions. This worry heightened their concern about supporting social research that might "embroil the Foundation in political controversies." Under such conditions, Alpert had to maneuver with great care.[13]

Help from NSF's general counsel enabled Alpert to determine what types of social science activities the agency could pursue, in light of its legislative origins and founding charter. Among other things, the agency could assess the state and needs of the social sciences. It could use them to examine the impact of science on economic development and the national welfare. It could use them to study scientific and technical manpower. It could also extend other agency activities to the social disciplines, including the award of research grants and fellowships. As the general counsel reminded NSF leaders, however, when it came to the social sciences Congress expected the agency "to exercise a fair amount of restraint in the use of this authority."[14]

In order to develop a viable policy framework, Alpert consulted widely. He spoke with leaders from the NSF, the large private foundations including Rockefeller, Carnegie, Ford, and Russell Sage, the SSRC, and disciplinary social science associations as well as with other prominent individuals in the social sciences, natural sciences, and government. From all sides, the main message to Alpert came across clearly: exercise caution.[15] More specifically, Henry Riecken, Alpert's successor at the NSF, noted that "Alpert was strongly advised by the friendlier members of the National Science Board, as well as by allies in the rest of the scientific community, to adopt a strategy of stressing the 'hard science' aspects of the social disciplines."[16]

Critical here is Riecken's point that stressing hard science comprised a strategy for funding social science in a manner that seemed unlikely to invite external criticism, that promised to respect the putative boundary between science and politics, and that Alpert could thus recommend to NSF leaders. Concerns presented by the psychologist and NSF staff member John Wilson indicated the salience of such issues. As Wilson saw things, focusing on the hard-science part of social science reinforced the vital distinction between social science and social action and thus helped to prevent the latter from contaminating the former. In a 1954 paper, he discussed the case of "action research" to illustrate the resulting problems when people failed to keep these two arenas—research and action—separate. In action research, he said, scientists often neglected "the basic issue of objectivity . . . a fundamental characteristic of science," and thus they could easily "confuse the social roles of the scientist and the social engineer." He found it especially problematic when scholars engaged in "social science to influence social change."[17] Wilson had previously worked under Waterman as the head of ONR's Personnel and Training Research unit, before Waterman recruited him for a post at the new agency. At the NSF Wilson headed the Psychobiology Program in the Biological and Medical Sciences Division from 1952 to 1955 and then became the division's assistant director from 1956 to 1961. Alpert therefore had good reasons for taking Wilson's concerns about separating social science from social action seriously. Indeed, Alpert, as shown below, would insist on this separation in NSF policy documents.

Alpert also took special note of Chester Barnard's prudent advice. As a board member and then president of the Rockefeller Foundation, Barnard was a main participant in the postwar controversy about the relationship of the social sciences to the problem of moral deficit. Barnard, as indicated in the previous chapter, spoke of a continuum of social research from soft to hard. He also proposed that the soft parts should not be neglected. Despite the inclination of many scientists to proclaim "an as yet undemonstrated unity of science," he perspicaciously observed that the "scientific literature and discussions among scientists of all kinds would suggest variations so great in the actual research procedures in different fields that these fields would have to be considered as different species, if not classes, of sciences." Furthermore, he argued it would be a mistake to insist that the social sciences should always take the natural sciences as the appropriate model.[18] Though Barnard's biographer William Scott claims that at the NSF Barnard underwent a "sudden conversion to orthodox positive science methods," Barnard, in fact, underwent no such conversion. Instead, Barnard advised Alpert to focus on hard-core research, meaning "the scientific aspects of social science research," because of the specific challenges involved in establishing a viable basis for NSF social science funding. Barnard's prudent advice also had special force because he now chaired NSF's governing board—Barnard had replaced Conant in this position.[19]

Against this complex background, in July of 1953 Alpert presented his superiors with a proposal to demarcate a realm of social science that could be separated from value-laden inquiry and social action. To ground this separation, he adapted Barnard's ideas as follows:

> The term "social sciences" covers a wide range of activities. . . . a continuum. At one end . . . lie the hard-core scientific studies . . . [these] include the use of experimental techniques, controlled experiments, laboratory studies, statistical and mathematical methods, and survey design techniques, the development of measurement devices and instruments such as standardized tests and scales, the empirical testing of hypotheses and concepts, and other characteristic features of scientific research. At the other end . . . lie the philosophical, ethical, and political studies and interpretations of human social conduct . . . [these involve] philosophical interpretations of social welfare, concern with the amelioration of social conditions and elimination of social problems, and similar considerations relating to social values and the good life.

After identifying this "wide range" of social science activities, Alpert then turned his attention to the hard-core end of the continuum, which he defined in terms of a universally valid scientific epistemology and methodology. He spoke of a "common core of logical organization, experimental procedures, mathematical and quantitative techniques, and conceptual cumulations" that established the "unity of science" and ensured the "objectivity" of scientific studies. Alpert also claimed hard-core research fulfilled "the highest standards of 'objectivity, verifiability, and generality.'"[20] As noted earlier, Bernard Berelson, Ford Foundation BSP director, also appealed to the principles of objectivity, verifiability, and generality and contrasted social science that embraced such principles with scholarship in a humanistic or philosophical vein.

Notwithstanding this move to restrict NSF support to the hard core, Alpert recognized that the agency's involvement with social research could still provoke public scrutiny, especially if the research dealt with socially sensitive issues. In order to address this threat, Alpert proposed that private patrons, particularly the enormous Ford Foundation, should provide the major source of social science "risk capital"—a position that Ford and other private foundations said they favored, though they also faced significant pressures to avoid certain controversial topics, especially during the McCarthy Era, as we saw in the previous chapter. Private funding, continued Alpert, should be responsible for "supporting the unorthodox, the unusual, and the 'big gamble,' as well as areas, like sex and politics, which can easily lead to public controversy." Alpert further argued that unlike the private foundations, whose programs tended to be "problem-oriented" and concerned with "social and political action," the NSF should concentrate on "the support and stimulation of basic research."[21]

With the task of presenting the social sciences in a nonthreatening manner nearly complete, Alpert still had to find a suitable space for them within NSF's structure. Since the existing divisional organization provided no obvious place, Alpert devised a creative solution by proposing that the agency concentrate on convergent research. This linguistic creation, like the terms "behavioral sciences" and "human resources research," could be useful in warding off potential criticism of the social sciences. But what exactly did convergent research consist of? In explaining this term, Alpert took his cue from Vannevar Bush, always a soothing reference point for NSF's natural science–oriented leaders. Taking a passage directly from Bush's postwar congressional testimony, Alpert said the agency should facilitate "effective integration and partnership between the natural and social sciences." In particular, the NSF should support research and fellowships that formed a bridge in terms of subject matter and methodology between the social and natural sciences. According to Alpert, good examples included research in "experimental social psychology, mathematical sociology, econometrics, operations research, information theory, communication theory, and similar fields."[22] All these fields attracted support from national security agencies, a point that may have encouraged NSF leaders to view Alpert's proposal here more favorably, though Alpert himself did not mention national security relevance as a consideration.

After Alpert presented his background study and policy recommendations, the decision about whether to go forward rested with those above him. His effort would soon pay off, though in a modest manner befitting the many limitations of those recommendations and the factors supporting those limitations, including the agency's natural-science orientation, its puny size and inadequate budgets, its commitment to an apolitical conception of basic scientific inquiry, and the palpable threat of conservative political hostility. As a result, Alpert's recommendations ignored the complexities of his own views concerning vital epistemological and methodological differences between the social and natural sciences as well as the need for social research oriented toward the realization of humanistic values such as human dignity and equality.

Surviving Conservative Threats via Protective Coloration

In the summer of 1954, the agency's leaders approved the establishment of a "modest, experimental program of research support." NSB minutes noted that this program had two branches and would be "cautiously developed," "based on continuous study and careful evaluation . . . over a three-year period."[23] One branch in the Biological and Medical Sciences Division had a Program for Anthropology and Related Sciences, which supported research in anthropology, functional archaeology, human ecology, demography, psycholinguistics, and experimental and quantitative social psychology. The other branch in the Division of Mathematics, Physical, and Engineering Sciences had a Program for the Socio-Physical Sciences that supported research in mathematical social science, human geography, economic engineering, and statistical design, as well as the history, philosophy, and sociology of science.[24] After being placed in charge of this new program, Alpert explained and promoted NSF funding opportunities throughout the scientific community. In doing so, he would leave aside any concerns he had about scientism in general and NSF's embrace of scientism as the path forward in particular. Meanwhile, NSF's carefully circumscribed program attracted some antipathy from conservative political interests, which, in turn, reinforced fears among agency leaders about becoming too closely associated with the social sciences.

Compared with their siblings in the natural sciences, the new convergent programs were toddlers. Total expenditures of the Socio-Physical Sciences Program amounted to $58,000 in 1955 and the same again in 1956, making it by far the smallest program within its division. The second-smallest program, in astronomy, had funds four to five times as large. Meanwhile, the Anthropology and Related Sciences Program, also the smallest program in its division, awarded research grants amounting to $52,000 in 1955 and $133,000 in 1956. The next-smallest in the division was the Developmental Biology Program, which awarded $156,000 in 1955 and $211,000 in 1956.[25]

Early successful project proposals for convergent research advertised their hard-core scientific orientation proudly, often by emphasizing a commitment to quantitative analysis. The following project titles were typical: Mathematical Theory of Economic Models and Related Topics; Research in Multiple Factor Analysis; Mathematical Techniques in Psychology; and Biometrical Study of Evolution.[26] By 1956 the NSF had also extended its predoctoral and postdoctoral fellowship programs to support convergent training and research.[27]

For NSF leaders, entering the contentious social science arena through convergent research made good sense, for many reasons. Supporting such research required a minimal amount of reorganization, as convergent programs resided under the administrative auspices of NSF's established natural-science divisions. In this arrangement, wary natural scientists oversaw the handling of social science grant and fellowship proposals, thereby providing these suspect newcomers with careful guidance and helping to identify, as Alpert put it, "projects of the highest scientific merit."[28] Placing the social sciences under a natural-science umbrella also promised to shield the NSF from political attack. Finally, this arrangement reinforced the agency's commitment to the unity-of-science viewpoint while respecting the reigning scientific pecking order.

From the social sciences' viewpoint, the pursuit of convergent research followed naturally from the underdog's position. Underdogs in many walks of life regularly adopt a strategy of "protective coloration" in their quest "to secure a share of power and position," as Henry Riecken noted. Unable to get what they want on their own, underdogs seek alliances with those who can help them obtain what they covet, a strategy of "allying one's cause with stronger others."[29]

Chapter 1 explained that during the 1940s, SSRC's social scientists employed this strategy by trying to catch a ride on the coattails of the natural scientists, as Talcott Parsons so revealingly stated. Their effort had largely failed at the time, but now, in the mid-1950s, the strategy of protective coloration had a better chance to succeed at the NSF.

Though Riecken did not mention it, the social sciences' search for protective coloration also raises a basic question about the wisdom of the unity-of-science position taken up by the NSF and other patrons. When used in a biological context, this strategy suggests not only protection for the weak but also disguise or deception. A leaf-eating insect might blend in with the foliage in order to avoid insect eaters. A bottom-dwelling fish might look like the seafloor in order to fool predators swimming above. But the insect does not turn into a plant. Nor does the bottom dweller become the seafloor. Nature's creatures might cleverly pretend to be what they are not, but successful creatures do not thereby transform themselves. So, in the case of the NSF, were the social sciences merely playing it safe, searching for protection from the more powerful and better-funded natural sciences? Or were all the sciences really of a similar nature? If one answered the latter question in the affirmative, then the strategy of playing it safe might well seem reasonable. But if one believed the social and natural sciences really had some important differences, concerning the ontology of their subject matters, their methodology, their epistemology, or their social relevance, then one might worry that the strategy of protective coloration would misrepresent their true character, distort their development, and undermine their scientific and social value. As for Alpert, he hoped that if the program of convergent research went smoothly, the social sciences would someday be able to leave the underdog's strategy behind.

But in the meantime Alpert would promote NSF's convergent programs to the academic world as part of the larger project of raising American social science from the morass of epistemological confusion, moral suspicion, and political harassment. In this promotional work, Alpert also took a critical step by presenting the cautious and limited character of NSF's initial social science efforts as the basis for progress in a broad sense. While doing so, Alpert relied on a tactic he had used to good effect before. Namely, he cited other well-known and respected figures from the social and natural sciences in support of his recommendations, thereby implying that the NSF and he—the agency's most important social science representative—simply agreed with a widespread consensus among leading scientists. Of course, had such a widespread consensus really existed, the social sciences would not have needed to work so hard to establish their scientific legitimacy to patrons like the NSF in the first place.

In numerous academic papers and talks Alpert identified the young agency as a leader of needed reforms throughout the social sciences. In a 1954

American Sociological Review article, Alpert said he agreed with the viewpoint of Elbridge Sibley, who had authored a major SSRC– and Carnegie Corporation–sponsored study on the recruitment and training of social scientists. Just as Sibley had said, Alpert now asserted that social scientists should formulate "hypotheses" that could be "tested and verified by experiments or systematic observations" and should develop more reliable and accurate predictions. In brief, social science had to "conform to descriptive analyses of science generally." Citing the amateur though highly influential historian of science and first NSB chairman James Conant, Alpert recommended that the social sciences "must progress toward firmer theoretical formulation and conceptualization." Agreeing with the sociologist and NSB member Charles Dollard, Alpert declared that social scientists needed to "become scientists in fact," by "dealing in fundamental theory which can be tested" and by "struggling to achieve laws and generalizations which will enable us to predict what men or groups of men will do or not do under stated conditions."[30] As Alpert elaborated in a later publication, this last notion implied that much like the engineer who "continually relies on the findings" of basic research in the physical sciences and mathematics, "practitioners of applied social science" needed "important theories and understandings of specific happenings" that would be provided by "basic research on the fundamental workings of human behavior."[31]

Alpert further indicated that by following these general principles of good science, social scientists would help people to distinguish social research from programs for social change, particularly those with leftist agendas. As we have seen, numerous social scientists and their supporters made this point during the early Cold War years. If by any chance the point remained unclear, Alpert specified that his social science peers should "eschew identification with social reform movements and welfare activities, and especially, the unfortunate phonetic relationship to socialism."[32] In this fashion, Alpert sought to stake out a safe space for the social sciences, while worrisome political developments provided him with a strong motivation to do so. The advent of NSF's convergent programs coincided with the cresting of McCarthyite attacks on un-American scientists, nuclear-weapon-data spies, and left-leaning social thinkers. Conservative attacks on social scientists and their patrons, including the large private foundations, were also reaching a feverish pitch.

In 1954 Alpert sent his assistant Bertha Rubinstein to the Reece Committee hearings—after working with John Wilson on an ONR-funded psychology project, Rubinstein had accepted his offer to join the NSF. At these hearings, Representative Reece, his loyal staff, and some social science witnesses expressed strong opposition to scientism and claimed the empirical quantifiers were leading the country toward moral ruin and a socialist nightmare.[33] In 1955 Representative Reece also mentioned the NSF by name in one of his tirades, indicating that some social research projects mentioned in an NSF report could never stand on their own if the "magic word [science or scientific]" were removed. Though the Tennessee legislator did not point his finger at any particular NSF-supported study, he did call attention to one project on prostitution in foreign countries and other social science projects that had federal funding from the Department of Agriculture and the Department of Health, Education, and Welfare.[34]

Shortly after Reece's comments, Alpert informed NSF director Waterman that "large-scale social science research programs initiated in government agencies" had often been "abolished, discontinued, restricted in scope, or budgetarily emasculated." Alpert's long list of such programs included many subjected to anti–New Deal and anticommunist attacks during the 1940s and early 1950s, including the OWI, the Department of Agriculture's Special Surveys Division, the Housing and Home Finance Agency's Housing Research Division, the U.S. Information Agency's basic research activities, as well as the air force's HRRI and its Inter-industry Economics programs. As impressive as this list already was, Alpert added that it included only "a handful of examples" selected "from an unhappily large number of significant social science programs of the Federal Government" that had "withered or died." The most recent case involved a congressional report recommending the elimination of the Department of Agriculture's support for projects on sensitive topics such as population dynamics, child-rearing practices, and differences in clothing between rural and urban people. The report judged such research "nonessential and non-productive," based on long-standing conservative hostilities that, as noted in chapter 1, had previously crippled Agriculture's social science planning efforts.[35]

In the very same month as Reece's remarks, John Teeter suggested the NSF itself was vulnerable to such hostilities. Remember that during the postwar NSF debate, Teeter, a conservative aide to Vannevar Bush, told Talcott Parsons that the social sciences should be satisfied with a coattail ride. Now writing to NSF director Waterman, Teeter claimed that "the large proportion of social studies in the last budget nearly wrecked the Foundation's reputation." "My sampling of congressional opinion," continued Teeter, "indicates a lot of pieces need to be mended."[36] After the agency announced that its fellowship program would include convergent fields, Teeter wrote to Waterman again, this time questioning whether the agency had the authority to support social science. He also warned Waterman that Congress might "frown by way of budget cuts on social science projects."[37] Since Teeter evidently disliked the social sciences and, as Daniel Kevles notes, he was "the kind of conservative who often confused liberalism with socialism," the accuracy of his claims about NSF social science funding nearly wrecking the agency's reputation may be questioned.[38] Still, Waterman could have easily checked with the relevant congressional appropriations committee about the alleged damage.

FIGURE 10. Harry Alpert. 1977. *Courtesy of Spencer Alpert.*

But, whether Teeter's claims were accurate or not, a cautious Waterman tried to reassure him, asserting that NSF's involvement with the social sciences remained within safe bounds. As Waterman pointed out, the decision to include convergent work in the fellowship program rested on "very careful consideration." Moreover, he told Teeter that the NSF did "very little" in the social sciences.[39]

Indeed, the agency's social science efforts through the mid-1950s had been and remained carefully circumscribed by design. Following the strategy of protective coloration first crafted by Alpert, the agency focused exclusively on the so-called hard-core end of the social research continuum and especially on what Alpert called convergent research. In this way, Alpert and NSF's natural science–oriented leaders hoped the agency would not be "accused of subsidizing a particular point of view," to borrow a phrase from the British political scientist and socialist Harold Laski, who had written about similar issues though in a different context two decades earlier. Designed to avoid partisan attacks and anticommunist pressures and also to uphold strong concerns within the agency about insulating science from politics and keeping social research separate from social action, NSF's early social science efforts thus rejected any interest in "value and ends" in favor of "material and methods."[40] Elsewhere, Alpert himself recognized the value of a much broader range of scholarly work. As indicated by the following statement of his from 1956, he also had a rich appreciation of the historically contingent character of alternative viewpoints about science and scientific method: "Science is pluralistic and dynamic, and, at different moments of history, one finds differing conception of scientific method. What is one generation's science may turn out to be the next generation's superstitions."[41] Yet Alpert's promotional efforts urged the nation's social scientists to embrace NSF's scientistic approach as their own.

Expansion within Restricted Boundaries

Starting in the mid-1950s, more congenial days for the social sciences and for the still rather small and struggling science agency arrived. The waning of McCarthyism, a moderate surge in predominantly liberal congressional support for the social sciences, and a heightened intensity in Cold War scientific competition all gave Harry Alpert and his successor Henry Riecken new opportunities to strengthen and expand NSF's social science program. At the same time, the agency's cautious natural science–oriented leadership ensured that such expansion would remain within narrow boundaries.

After the downfall of Senator McCarthy at mid-decade, domestic Red-baiting became less virulent, and conservative assaults on the social sciences and their patrons diminished. Additional congressional investigations of private patrons had seemed all but certain at the conclusion of the Reece Committee's work. However, no major investigation took place. Even the 1955 Wichita Jury Study investigation focused mainly on a single BSP-sponsored project. As it turned out, then, BSP's demise occurred just as the political climate became less repressive.

In addition, following the launch of *Sputnik 1* in 1957, heightened fear that America lagged dangerously behind in the Cold War struggle for scientific and technological leadership also provided an opening for the social sciences. In the post-*Sputnik* panic, American political and scientific leaders launched ambitious new science initiatives including, among other highlights, the President's Science Advisory Committee (PSAC), the National Aeronautics and Space Agency, and the National Defense Education Act. National interest in basic research soared as well. Even President Eisenhower declared America had a "long-term concern for even greater concentration on basic research," while director Waterman forwarded an NSF report called "Basic Research—a National Resource" to the White House.[42] Rising federal science budgets followed. From 1957 to 1961, federal R&D support doubled and support for basic research tripled. NSF budgets increased more than 250 percent, making it easier for the agency's leaders to allocate additional dollars to the social sciences.[43]

Meanwhile, positive congressional interest in using social science to solve social problems rose, with some liberal Democrats and a few moderate Republicans expressing direct interest in NSF's efforts. Jacob Javits, a New York Republican senator who promoted school desegregation and civil rights, placed a resolution from the American Sociological Society requesting an expansion of NSF's social science program into the *Congressional Record*. Wayne Morse, a maverick senator from Oregon who started his political career as a Republican with Progressive roots before switching to the Democratic Party, cited increasing congressional concern about the agency's "slowness and excessive caution" when it came to funding social science.[44]

This pro–social science legislative contingent included Hubert Humphrey, a Democratic senator from Minnesota. As an undergraduate student at the University of Minnesota, Humphrey aced nineteen political science courses and became good friends with one of his political science professors, Evron Kirkpatrick. After completing a master's degree at Louisiana State University for a thesis that applauded the New Deal and praised President Roosevelt as a pragmatic experimentalist, Humphrey planned to pursue doctoral studies. But with Kirkpatrick's guidance, Humphrey moved into Democratic politics instead, becoming a Minnesota senator and later the vice president under Lyndon Johnson, before making a run for the presidency in 1968. Following *Sputnik 1*, Humphrey inserted the Miller Report into the *Congressional Record*. As discussed in chapter 2, this report was inspired by concerns raised by Vice President Nixon. It warned that the communists could acquire a dangerous advantage in the social sciences. And it thus urged increased national funding for them.[45]

Some congressional committees expressed interest in NSF's social science programs as well, including the Senate Subcommittee to Investigate Juvenile Delinquency chaired by Tennessee Democrat Estes Kefauver. An independent-minded, liberal internationalist who strongly supported civil liberties but also racial segregation up until the 1954 *Brown* decision, Kefauver became a formidable political presence during the 1950s and defeated John F. Kennedy to gain the vice presidential spot on the 1956 Democratic ticket alongside Adlai Stevenson. Kefauver's Senate subcommittee focused on a subject that crackled with national interest and inspired research by many scholars, such as Paul Lazarsfeld, Robert Merton, the Gluecks, and Richard Cloward and Lloyd Ohlin, as noted in chapter 3. Patrons of this research included the Ford Foundation and federal agencies such as the National Institute of Mental Health and the Children's Bureau.[46]

During legislative hearings, Kefauver's subcommittee heard conflicting views about the causes of and cures for delinquency from Lazarsfeld and other experts, thus suggesting the need for additional research on topics such as the effects of mass media on American youth. According to a common fear, the mass media contributed to a rift between parents and children and thus disrupted the transmission of decent values to the younger generation. But the prevailing theories did not agree on the causes or remedies. In a 1957 report, the subcommittee emphasized the need to overcome "serious gaps" in the available "knowledge of the fundamental mechanisms of human behavior in society," called for more research, and recommended additional NSF funding.[47] Seeing a golden opportunity, Alpert fed Kefauver materials about federal social research funding and its limitations.[48]

Kefauver's subcommittee did not demand that the NSF support applied studies though. Instead, the subcommittee emphasized the value of additional funding for "fundamental" research, involving hypothesis testing, experimental verification, and systematic observations, for such work promised to help society move from "the realm of speculation and subjective opinion . . . to the realm of assured and demonstrated fact." Nevertheless, the subcommittee also understood that such research would play a critical role in the "attack upon social problems" by establishing the basis for "rational, effective, action programs."[49] In an excellent book about juvenile delinquency during the 1950s, James Gilbert claims that pressure from Kefauver's subcommittee and other social science–friendly legislators led the NSF to venture into "this area of basic investigation."[50] However, the NSF had already been funding a modest amount of basic social research. More importantly, the subcommittee's recommendations provoked resistance from NSF leaders, though Gilbert's account does not explore this point.

After all, the subcommittee's enthusiasm for linking fundamental research to action programs did not fit comfortably with NSF's limited program. As director Waterman saw it, the topic of juvenile delinquency lent itself mainly to applied studies, which other agencies should support, such as the Department of Health, Education and Welfare, but not the NSF. Regarding this field of study, Waterman explained that his agency would "in the usual way" consider funding proposals for "good basic research." The NSB arrived at the same position, reinforcing existing limitations on the agency's social science efforts by reasserting the separation of basic studies from value-laden inquiry and policy-oriented work.[51]

Still, this positive congressional interest delighted Alpert, who now placed the case for expansion before Waterman. As well as anybody, Alpert understood that "the fate of public support of social science research, particularly at the federal level," depended on "the fundamental attitudes of key congressmen."[52] While he agreed that the agency should continue to "proceed cautiously," he also proposed that it should consolidate, upgrade, and expand its social science efforts beyond the convergent fields, yet not so far as to include "applied and mission-related social science research." Alpert told Waterman that three years of experience had proven that the NSF, by relying on "a careful and thorough selection procedure," could "identify significant and scientifically meritorious projects in selected social science areas."[53]

Soon after Alpert's modest recommendations, the agency proceeded with some enhancements, including the appointment of an eminent new social science leader. In 1958 Riecken replaced Alpert, who went to the University of Oregon as Graduate School dean and later became provost. Riecken's impressive résumé included wartime experience in military aviation personnel selection and training, a Harvard doctorate, a University of Minnesota professorship, research grants from major patrons including the ONR, Carnegie Corporation, and Ford Foundation, and a position as adviser to the Ford Foundation's BSP. In the late 1950s, Riecken also served on DOD's Steering Group that oversaw a major review of military social science programs and involved the Smithsonian Institution's group led by Charles Bray. In 1965, Riecken would leave the NSF for the SSRC, where he served as vice president and then president—replacing Pendleton Herring, who completed a remarkable twenty-year tenure in that position.[54]

As Riecken explained, his "professional identification with experimental social psychology fitted well with the generally quantitative and empirical orientation of the NSF program."[55] A pioneer in the scientific evaluation of social programs, Riecken agreed with the view that the social sciences had "only recently become differentiated from ethical and moral philosophy." Until recently, they had not been "empirically grounded sciences" and suffered from "a cacophony of conflicting views." Fortunately, scholars "practicing social *science* rather than social *opinionating*" had paved the way for great progress in the past quarter century, a difference Riecken deemed "at least as great as that between chemistry and alchemy." He thus felt quite comfortable working with the established scientific order at the NSF and in the national science system more generally—or the "gentleman's club," to borrow the words of longtime NSF social science assistant Bertha Rubinstein.[56]

The NSF also implemented a series of upgrades in the organizational status of its social science efforts. Replacing its two separate convergent programs, the agency first established a single social science program and, shortly thereafter, elevated its status to office. Starting in 1961, the social sciences even acquired their own Division—which surely qualifies as a minor miracle. Regarding this landmark event, a *New York Times* article reported that "the social sciences received a long-sought recognition today by being elevated to the status of the physical and biological sciences in the government's program of basic research." As this article noted, director Waterman himself emphasized "the symbolic importance of the action in placing social sciences on an equal footing with the more traditional fields."[57]

These upgrades accompanied reorganization at another level, as the agency replaced its convergent research programs with disciplinary-based ones. With the development of new programs for anthropology, sociology, and economics, proposals in these disciplines no longer had to undergo screening and review by natural scientists, though final approval of agency grants still rested with its natural science–oriented governing board. These three disciplinary programs and a fourth one for history and philosophy of science became the core of the agency's new Social Science Division.[58] Though the NSF had no separate psychology program, biologically oriented psychologists could apply for support to the Psychobiology program in the Biological and Medical Sciences Division, and social psychologists could apply to the Sociology Program.

Under these conditions, social science–friendly legislators began calling for greater NSF support and complaining about what they took to be a lopsided emphasis on the natural sciences. Senator Wayne Morse placed a letter from the anthropologist Luther S. Cressman into the *Congressional Record* in which Cressman pointed out that for every federal dollar spent on research in the mid-1950s, the physical sciences received eighty-seven cents, the life sciences eleven cents, and the social sciences merely two cents.[59] The distribution of NSF funds likewise remained heavily skewed, added Charles O. Porter, a Democratic representative from Oregon. Concerned about the growing "crisis" in U.S.–Latin American relations, Porter charged that journalists and "so-called experts" in Latin American affairs often adopted "a cynical attitude toward democracy in the countries below the Rio Grande," while the United States itself had often placed its "power and prestige . . . on the side of the dictatorships."[60] Drawing on a vision of modernization that posited a close association between social science, economic progress, and the development of democracy, Porter anticipated that greater federal support for social research on Latin American affairs would contribute to a more enlightened American foreign policy. As for specific dollar amounts, Porter observed that NSF's 1959 research budget estimate of $40 million represented a nearly 250 percent increase over the previous year's actual budget, but the $600,000 earmarked for the social sciences represented only a 50 percent increase. Hoping to prevent the social sciences from losing additional ground in relation to the natural sciences, Porter introduced an amendment to the agency's appropriations bill that would double social science funding. However, this amendment failed, partly because the NSF itself had not requested the change.[61]

Limited social science representation in NSF's leadership provoked additional congressional complaints. At one point during the 1945 congressional hearings on NSF legislation, the sociologist William Ogburn worried out loud that "if you put 1 social scientist in with, say, 9 or 10 natural scientists, and if you have a director who is a natural scientist, they will forget all about the social scientist being there and they will give him very scant attention."[62] In 1958 the Senate Committee on Labor and Public Welfare sought to alter the reality Ogburn had anticipated. Hoping that the need to fill nine NSB openings that year would benefit the social sciences, the Senate committee provided President Eisenhower with a list of seven influential figures for consideration, including Donald Young, Dael Wolfle, John Gardner, and Clyde Kluckhohn. Apparently, the SSRC had suggested these individuals to the committee. At the time, only two NSB members had advanced social science training, Frederick Middlebush and Charles Dollard, and Dollard's term was ending.[63] Yet not one of those scholars and social science administrators received an invitation to join NSF's governing board. In fact, the president, who by law had responsibility for nominating NSB members, did not even suggest a single social scientist. Representative Porter thus accused Eisenhower of weakening "the national position of the social sciences" and predicted that "the President's action" would be "a blow" to NSF's "small but promising" social science program.[64]

Even Riecken—hardly the boat rocker—complained to director Waterman about inadequate budgets and relative neglect. Alpert had often "found it necessary to pare all requests to the very minimum," noted Riecken. Now and then Alpert had even decided to decline "a highly meritorious proposal whose budget, time span, or sphere of activity could not be reduced without propounding an absurdity." After Riecken took over, NSF's average social science grant remained "appreciably smaller" than its average natural science grant. The average duration of social science grants remained "shorter" as well.[65]

But the agency continued to handle the social sciences cautiously. In a nod to mounting congressional interest, NSF's *1959 Annual Report* acknowledged that "the intellectual, economic, and social strength of our nation requires a vigorous approach to social problems, with scientific techniques of study making their maximum contribution."[66] Yet the agency still insisted on defining the scope of its interest in a way that distinguished basic social science from social action and social problem solving. Explication of this restriction closely echoed Alpert's early formulation. Director Waterman told the NSB that it was "proper and desirable for the NSF to support basic research in the social sciences, where such research meets the usual scientific criteria of objectivity, verifiability and generality."[67] The NSB itself approved a recommendation from the agency's external Advisory Committee for Social Sciences to support only "fundamental research that was not concerned with practical applications."[68] Similarly, upon the creation of the Social Science Division in 1961, Waterman reiterated that the agency's interest lay in "the study of basic problems in the social sciences and the development of research techniques," but not "applied craftsmanship in social affairs or in contemporary problems."[69]

In sum, during the late 1950s and early 1960s, NSF's social science efforts expanded through the appointment of Riecken and important upgrades in organizational status and funding, culminating in the establishment of a separate Social Science Division, which no doubt marked a major milestone. A thaw in the political climate, growing congressional support, and the post-Sputnik boom in scientific activity enabled such enhancements. Still, throughout the 1950s NSF funding for the social sciences represented only 1.4 percent of the agency's total support for scientific research.[70] When friendly and mainly Democratic legislators such as Senator Kefauver and Representative Porter called for a more robust social science program, NSF leadership reasserted the limits of the agency's interest, namely, its support only for basic research and not applied work, policy studies, or normative inquiry. Thus the social sciences' "breakout from convergence" amounted to a "modest walk, not a dash," as the sociologist Otto Larsen has written in his insider's history of the agency.[71] Later sections of this chapter will explore more closely why critics both inside and outside the agency found it so difficult to mount successful challenges to the agency's cautious approach to the social sciences. However, before turning to those challenges, this next section will examine NSF's increasing importance for the social sciences during these years.

Dollars, Research, and Scientific Status

Despite the frustrations experienced by those who wanted the agency to become more deeply involved with the social sciences, obtaining a foothold at the NSF still represented a major step for these sciences in a number of respects. By 1955 the NSF was already the fourth-largest federal patron for extramural, unclassified social research. In 1960, the NSF provided $4.2 million and was still in fourth position, while the Defense Department provided $16.9 million, Health, Education and Welfare provided $17.4 million, and Agriculture $17.5 million.[72] Furthermore, while federally funded extramural social science had "in very large measure" an applied character, "related to the specific missions of the agencies supporting the research," the NSF, as Alpert noted, focused exclusively on basic research.[73] In fact, for many fields of basic study, the NSF quickly became the major source of extra-university funding, and sometimes the only source. Regarding federal support of "basic research in such areas as anthropology, economics, demography, and the history and philosophy of science," the NSF stood alone or nearly so as early as 1957.[74] In some cases the agency funded work of interest to other major patrons including Ford and the military, though the NSF also retained a distinctive character through its more exclusive focus on scholarship rather than action-oriented research. In all cases the young agency remained wedded to a scientistic orientation, as seen in the agency's engagement with the individual disciplines, its funding for particular projects, and its promotion of broader science policy principles supporting the unity of the sciences and the separation of social science from politics, social action, and other "softer" forms of social inquiry.

In 1960 the NSF provided 60 percent of all federal dollars for sponsored research in anthropology, while the National Institutes of Health provided 35 percent and the Smithsonian Institution 5 percent.[75] NSF's initial foray into this field focused on biological anthropology and physical archaeology. Both had popular appeal and close ties to the natural sciences. In Riecken's words, NSF's approach to archaeology concentrated on "scientific studies of human prehistory." Its first two grants supported research by Harvard's Gordon Wiley on prehistoric settlement patterns in the Mayan territory and Chicago's Robert Broadwood on human populations in the Fertile Crescent. Other successful project proposals focused on topics such as the late prehistory of northern Arizona and pre-Columbian culture. Meanwhile, Riecken noted that "classical archaeology, with its historical and humanistic emphasis," remained ineligible for support.[76]

By providing selective funding, the NSF also contributed to the ascendancy of "social science–archaeology," or "the new American archaeology." Lewis Binford, a leading advocate of this approach, argued that archaeologists had mainly contributed to "explication" not "explanation." But within the "scientific frame of reference," scholars focused on explanation, involving "the demonstration of a constant articulation of variables within a system and the measurement of the concomitant variability among the variables within the system." Accordingly, the new American archaeology had little to do with research that viewed data "particularistically." This new program thus reflected a broader midcentury movement within anthropology that emphasized the analysis of general cultural processes and evolutionary change in cultural systems, rather than the careful description of particular prehistoric sites, specific cultural facts, and unique historical conditions—a style of work closely associated with Franz Boas and his students.[77]

NSF's anthropology program supported work in demography as well, a field undergoing rapid development marked by quantitative analysis, predictive modeling, and Cold War relevance. Alpert reported that NSF's first grant in demography supported a Durkheimian-inspired study that aimed to develop "a unifying empirical principle for the analysis of variations in suicide rates." Another grant went to a study of "migration differentials." Later, in an internal NSF report to director Waterman, Riecken emphasized that demography lent itself "readily to quantification" and had recently acquired "statistical sophistication."[78] With extensive funding from private foundations including Ford, the field's quantitative orientation and predictive aims also made it relevant to the control of population growth in developing nations. Accordingly, a number of leading demographers and their patrons became active in establishing population policies designed to further American Cold War objectives in the developing world. However, this point about demography's strategic relevance did not fit with NSF's orientation and indeed did not appear in the reports from Alpert and Riecken.

Within sociology, those working on demography, quantitative and experimental social psychology, and human ecology first became eligible for NSF support through its Anthropology and Related Science Program. After the agency dropped the convergent criteria, other sociological specialties also became eligible. One early NSF grant encouraged research on the experimental and mathematical analysis of choice behavior, part of the agency's interest in the decision sciences. The agency also continued to emphasize methodological work aimed at establishing greater scientific rigor, through experimental studies, mathematical techniques, and computer simulations of social processes. Riecken emphasized in particular that "computers" allowed social scientists "to manipulate complex systems and to understand better their systemic properties."[79] Though Riecken did not say so, the military provided extensive support for computer simulation of social systems as well.

Though psychologists did not have their own NSF program, the agency welcomed psychological studies with a strong life sciences component and a quantitative, experimental orientation. One large grant provided $120,000 to the Yerkes Laboratory of Primate Biology in Florida, one of the world's leading centers for primatology and named after the self-proclaimed psychobiologist Robert Yerkes. During the postwar NSF debate, Yerkes had asserted that traditional demarcations between the social and natural sciences had little meaning. NSF's initial focus on convergent research and its effort to facilitate psychology's contact with the life sciences followed, at least indirectly, from Yerkes's position. Meanwhile, NSF's sociology program supported experimental social psychology, including research on the mathematics of imperfect discrimination and the landmark experimental studies on cognitive dissonance carried out by Stanford's Leon Festinger.[80] Thus, as historian of psychology James Capshew has written, NSF funding indicated that "psychology, or at least segments of it, belonged in the circle of the natural sciences."[81]

After economists first obtained support for mathematical economics and econometrics through the Socio-Physical Science Program, NSF's new economics program in the postconvergent period continued to fund the "most heavily mathematized" parts of their discipline, including the two subfields above as well as work in economic geography, economic productivity, game theory, input-output models, and some pioneering studies in laboratory experiments.[82] A typical NSF grant went to the National Bureau of Economic Research (NBER) for computer-based statistical analysis of business-cycle data. As Riecken described it, this project would help to develop "a new analytic tool that could vastly increase the rapidity and exactness of economic analyses." However, Riecken, always careful to separate NSF-supported studies from applied work, added that this project aimed only to devise and make available "more competent methods for analytic work," not to "develop economic policies."[83] This grant also comprised a fitting memento to the longtime NBER director Wesley Mitchell, who during the NSF debate had presented SSRC's argument for inclusion based on the unity of the sciences and had often made a point of separating value-neutral economic science from value-laden policy making. On the flip side, scholars looking for support to study the theory and practice of economic justice had no chance of securing NSF funding.

The three fields of history, philosophy, and sociology of science also found a home at the NSF, which may seem strange at first, especially the history and the philosophy of science parts because of their close association with the humanities.[84] But certain features of all three fields resonated with NSF's emphasis on keeping science insulated from politics and on placing the social sciences within a wider scientific enterprise led by the natural sciences. First, significant developments in these fields during the 1940s and 1950s suggested it was both possible and desirable to separate the "internal" development and the intellectual content of scientific work from "external" influences such as religion, politics, and patronage. Thinking about science in this way resonated with the common distinction at the time between the context of scientific discovery, which might involve all sorts of external influences, and the context of justification, which supposedly involved only internal scientific judgments about empirical evidence and logical consistency and thus not social pressures or ideological biases for or against some particular scientific claim or theory.[85] Second, in midcentury America these three fields largely opposed Marxist accounts of the class-based character of scientific development and its historical role in the eventual overthrow of the capitalist order.[86]

Third, these fields all presented science as a single, unified, and progressive enterprise that, in addition, had a special affinity with liberal democracy. Philosophers of science including Princeton's Carl Hempel and the European émigré Karl Popper concentrated on finding universally valid rules of scientific inquiry, methodology, and explanation. Popper also famously defended science as critical rational inquiry and then presented the scientific community as a model for the "open," liberal, and democratic society, directly opposed to the "closed," illiberal, and authoritarian alternative based on a pseudoscientific, Marxist ideology and other "sweeping historical prophecies."[87] As for historians of science, they emphasized a few central developments in the establishment of modern science and the progress of Western civilization. Works on the seventeenth-century Scientific Revolution by Harvard's I. B. Cohen and other scholars presented an impressive picture of scientific advance founded mainly on the efforts of a few key individuals from the physical sciences (i.e., Copernicus, Galileo, and Newton) who had overcome common superstitions, philosophical dogmas, and religious authorities in their effort to make quantitative analysis, empirical inquiry, and experimental method the hallmarks of modern scientific investigation.[88] Led by the outstanding Columbia scholar Robert Merton and his students, sociologists of science proposed that social norms governing scientific activity functioned to ensure scientific progress, claimed that democracy and science enjoyed mutually supportive relations, and warned that heavy-handed political controls could undermine scientific advance.[89]

In line with NSF's general orientation and priorities, its support for history, philosophy, and sociology of science concentrated overwhelmingly on studies concerning the natural sciences. In the early years, NSF's physics program supported work on the philosophical foundations of physics, while its mathematics program funded research on the logical basis of probability theory. The history, philosophy, and sociology of science also received support from NSF's convergent programs and then from the agency's more independent, unified, and disciplinary-oriented social science effort. One characteristic NSF-funded project, described in an internal document from 1960, focused on the uncertainty principle in twentieth-century physics. Another project examined structural theory in organic chemistry.[90] While funding for studies about the social sciences remained minimal, the NSF supported some scholarship on methodological similarities between the social and natural sciences. In the late 1950s, for example, the agency joined the SSRC in sponsoring a conference that led to an edited volume of essays called *Quantification: A History of the Meaning of Measurement in the Natural and Social Sciences* (1961).[91]

Beyond providing funding for projects marked by a dogged emphasis on scientific rigor in the ways described above, the agency's growing stature as a patron of American science made it increasingly valuable to the social sciences in symbolic ways. Director Waterman and NSB members claimed the agency stood for the promotion of the best science by the best scientists. Right from the start, the agency placed "scientific merit" at the top of its list of criteria for judging research proposals. Throughout the nation scientists and science administrators commonly believed, as former NSF historian George Mazuzan has written, that at the NSF "Good Science Gets Funded."[92] In addition, the NSF had a special relationship with universities, as university presidents, other high-level academic administrators, and university scientists dominated the agency's leadership while university scientists dominated its science advisory boards. In Roger Geiger's words, the agency served "first and foremost" as "a patron of the research universities."[93] Against this background, the NSF acquired an increasingly important role in judging the scientific merits of academic social science research projects and their investigators.

Furthermore, NSF's allegiance to the principles of intellectual freedom and scientific autonomy reinforced the notion that extra-university funding did not need to compromise the scholarly integrity of academic social science. Echoing the science policy principles in Bush's *SEF*, director Waterman declared that "it is incompatible with the nature of basic research . . . to chart its direction and progress." Thus the agency would not "endeavor to 'master-mind' basic research."[94] The agency's decision to rely on a system of peer review for judging proposals and to use grants for funding research, rather than contracts, reflected a strong commitment to advancing scientific knowledge. Because contracts specified the nature of the research product in greater detail, they seemed more appropriate for mission-oriented agencies. In this context, the NSF promised to give social science grant recipients control over their research, though the agency still demanded that researchers meet its hard-core funding criteria.

At the decade's end, two influential scholar-administrators, the psychologist Dael Wolfle and the political scientist Pendelton Herring, commended the agency for advancing social science along progressive lines. Wolfle had joined OSRD's Applied Psychology Panel during World War Two, served as editor of the APA's flagship journal *American Psychologist* from 1946 to 1950, and became a major figure in government-psychology relations during the Cold War years. In a 1960 *Science* editorial, Wolfle noted that he agreed with NSF's emphasis on supporting basic social research that met "high standards of conceptual and methodological rigor." By doing so, the agency helped to remove "any remaining fears that . . . social science is merely a cloak for action on important but sometimes controversial social issues."[95] Also in 1960 Herring, still the SSRC president and also the head of NSF's external social science advisory panel, wrote a favorable letter to the agency's governing board. Remember that in his Reece Committee testimony, Herring had combined a commitment to nonpartisan, empirical social inquiry with a spirited attack on Marxist dogma and pseudoscience. Now, and even though the agency had not yet identified Herring's own discipline (political science) as being eligible for funding, he claimed "support for basic social science research by the NSF is really crucial." With the demise of Ford's BSP, the nation's social scientists had "no comparable resource."[96]

To be sure, the gradual expansion of NSF's efforts during the 1950s had brought notable increases in funding for psychological, sociological, economic, and anthropological studies as well as for work in the history, philosophy, and sociology of science. In light of NSF's special focus on basic science, its good relations with the nation's research universities, and its commitment to scientific autonomy and peer review, the agency's support for the social sciences also had deep symbolic importance in midcentury America. The commentaries by Wolfle and Herring pointed to the agency's crucial role in furthering the scientific credentials of the social sciences.

Yet rigid limitations in the intellectual scope of NSF's social science program, the program's comparatively small size, and the marginal representation of social scientists within the natural-science-oriented agency also remained notable, though the positive assessments by Wolfle and Herring overlooked these points. Neither did Wolfle or Herring focus on the fact that restrictive funding criteria rendered large segments of social research that had an action, policy, or normative orientation ineligible for NSF support. During the second half of the 1950s, a group of legislators tried to push the agency to broaden and deepen its social science commitments, as discussed before. Policy-making dynamics within the agency also frustrated a few NSF insiders who valued a more expansive approach, as explained in the next section.

Worse Than Pandora's Box

NSF's published documents presented an agency moving slowly but steadily toward greater social science support within a consistent framework committed to the unity of the sciences and resting on the sharp separation of hard-core social research from value-laden humanistic inquiry, social action, and public policy. But inside the agency, differences of opinion about the nature and value of the social sciences sometimes provoked strong criticisms and reform efforts. Some elite natural science administrators and NSB members with close ties to Vannevar Bush continued to suggest social science did not fit well with the agency's mission, priorities, and structure. From a very different perspective, Father Theodore Hesburgh, another NSB member with a theological orientation and an appreciation of humanistic social inquiry, expressed interest in moving beyond the agency's narrow scientistic approach. The resulting conflict reached its peak intensity circa 1958 and involved the work of a special NSF committee established to deal with the controversial social sciences.

The year following the *Sputnik 1* launch, the NSB appointed a Committee on Social Sciences to assess past policy and make recommendations for the future. This committee, which bridged the employment tenure of Alpert and Riecken, included former Harvard geologist Donald McLaughlin, Rockefeller University vice president Douglas Whitaker, Frederick Middlebush (still the only NSB member with advanced social science training), and the new committee's chairman, Father Hesburgh. An energetic man with a broad intellect, Hesburgh obtained national and international prominence for his work in higher education and the Catholic Church. Since his childhood, Hesburgh had wanted to be a priest. He achieved this ambition. He also became a college professor at the University of Notre Dame and then its president. During the third quarter of the twentieth century, Hesburgh served on numerous national committees concerned with higher education. He became counselor to four popes and six American presidents. The president of the International Federation of Catholic Universities as well, Hesburgh assumed a leading role among Catholic educators aiming to break out of a largely self-imposed Catholic intellectual ghetto.[97]

The manner in which Hesburgh joined the NSB foreshadowed the challenge he would raise to the agency's social science effort. In 1954 President Eisenhower expressed interest in nominating Hesburgh for a position on the agency's governing board. At first Hesburgh hesitated, noting that he knew little about science. However, Eisenhower explained he wanted to have Hesburgh's humanistic and moral perspective present at the agency. The previous year, in his first inaugural presidential address, Eisenhower warned the nation about science's terribly destructive potential: as the "forces of good and evil" stood "armed and opposed as rarely before in history," science seemed to offer "as its final gift, the power to erase human life from the planet." Hoping to avoid impending doom, Eisenhower urged Americans to reaffirm their "faith in the deathless dignity of man, governed by eternal moral and natural laws."[98]

Hesburgh himself believed science should advance human welfare but also remain subordinate to values that lie beyond science itself. Science, he stated, needed to maintain "the balance and the supremacy of those older and better treasures of mankind—the humanistic values." Science becomes "meaningful and fruitful . . . only when it reckons with . . . deeper spiritual values . . . and these values come from outside the realm of physical science."[99] In the post-*Sputnik* period, Hesburgh feared the nation would focus too exclusively on science, in the process leaving behind more fundamental humanistic studies, including foreign languages, literature, history, and the arts.

Hesburgh and Eisenhower also shared a deep skepticism about quantitatively rigorous but spiritually impoverished social inquiry. Eisenhower, who had been president of Columbia University in the early 1950s, saw the value of higher education not in terms of producing specialized knowledge but in providing advanced general education for citizenship. In one episode Eisenhower called a halt to a social research project involving Robert Merton and other scholars who were analyzing mail to Eisenhower from the public urging him to run for president. According to Joan Goldhamer's examination of this episode, Eisenhower may have been afraid of provoking public outrage, should it become known that he had allowed scholars to examine his mail.[100] But regardless of Eisenhower's motivation in this particular case, sociologists chafed at Eisenhower's more general ambivalence toward their profession. C. Wright Mills quipped that while the nation's first president, George Washington, had read Voltaire's *Letters* and John Locke's *Essay Concerning Human Understanding*, Eisenhower read "cowboy tales and detective stories."[101] Most relevant to the present analysis, Eisenhower's religious faith together with his rejection of the communist enemy's "atheistic philosophy of materialism" led him to challenge the general tendency to define "truth solely in the narrow terms of mere fact or statistic or mathematical equation."[102]

Similar concerns informed Hesburgh's effort to turn Notre Dame into the greatest Catholic university in the world, a "Catholic Princeton." One of Notre Dame's previous presidents had visited the university library to dispose of offensive books, including sociology texts.[103] Though Hesburgh did not approve of that practice, he did find the amoral and naturalistic presuppositions in modern social science troubling. In his view, spiritual concerns should be "the apex of all university learning." Consequently, "the ultimate answers" sought by scholars needed to "impregnate the whole field of university teaching and learning." Thus, for example, the discipline of economics "does not exist for itself, any more than money exists for itself. All things are seen in relation to man's last end and first beginning which is God." Hesburgh thus set out to strengthen the social sciences at Notre Dame, but along lines that admitted a direct engagement with spiritual and humanistic values. Toward this end, he facilitated the hiring of the Catholic political scientist Jerome Kerwin from the University of Chicago.[104]

In a background paper prepared for Hesburgh's committee, Harry Alpert explained how NSF's approach to the social sciences had been intentionally restrictive. The agency had defined its program "so as to omit . . . controversial areas," such as "sex, religion, race, and politics." Hoping to avoid public criticism, the agency would not fund social research in "controversial fields no matter how significant such research might be." Its focus on hard-core research meant leaving aside those fields "concerned with considerations of ethics, welfare, and philosophical interpretations of man's destiny" as well.[105] Hesburgh agreed with Alpert that "the program thus far . . . had been at least a safe program, and a conservative one."[106] After hearing that the NSF did not support research on sex, race, religion, and politics, Hesburgh also hinted at his critical perspective by declaring that "that doesn't leave much."[107]

In June of 1958 Hesburgh presented his committee's policy recommendations to expand, strengthen, and add diversity to the agency's program. First, he proposed upgrading the organizational standing of NSF's social science effort from program to office, with an eye toward establishing a social science division in the near future. Second, he recommended adherence to the "same general philosophy" followed in the past. Third, he advocated expanding the agency's social science advisory committee in order to include "all of the various points of view and philosophies represented in social science" and to ensure that all "the various scientific disciplines in the social sciences have adequate representation."[108] Despite the second recommendation's conservative tone, the third one implied a major departure from past policy.

Hesburgh added that the committee members held "opinions far to each side," though he did not provide names or details. He simply noted that besides himself, some members believed NSF's program "should be considerably broadened beyond the present policy statement." However, other committee members took "a dim view of social science activity generally, unless very tightly restricted to its 'scientific' aspects."[109]

This dim view also surfaced at a governing board meeting, with some members from the nation's natural-science elite once again expressing antipathy toward social science. According to an NSB document, one unidentified member wanted to strike out the offending passage about "all of the various points of view" being "represented." Julius Stratton, an electrical engineer and at the time MIT's acting president, joined additional, unnamed board members in asserting that the agency should "exercise extreme caution" because expanding its social science program "would change the character of the Foundation and might necessitate a change in the composition of the Board," presumably by adding more social scientists. Furthermore, Stratton reminded his colleagues that lingering congressional suspicion of the social sciences could result in negative repercussions, such as cuts in NSF's "regular research budget."[110]

The mathematician Warren Weaver also voiced opposition. Weaver had achieved renown for his leadership of the Rockefeller Foundation's Natural Science Division (1932–1952) and for encouraging physicists and mathematicians to turn their attention to basic problems in biology. During World War Two he chaired OSRD's Fire Control Section and Applied Mathematics Panel. In 1947, he led the New York conference to recruit social scientists for RAND. So Weaver did not dismiss all types of social science nor their relevance to national security matters. However, he favored mathematical forms of social research. Furthermore, when he spoke of "science," he generally meant "of course, real science, basic science, such as physics, chemistry, biology, mathematics, and astronomy," as he wrote in his autobiography.[111] Regarding the proposals from Hesburgh's committee specifically, Weaver asked if the NSB really wanted to support studies beyond the convergent fields. In fact, the board had already decided to do so a few years ago, as Waterman pointed out.[112] Apparently Weaver did not know about this decision, and presumably he did not approve.

Adding to the controversy, the geologist on Hesburgh's committee, though he had earlier moved to accept the proposals to expand NSF's social science program, now backtracked. In light of the concerns raised by Stratton, Weaver, and others, McLaughlin suggested the committee should reconsider the matter.[113]

Following this stormy meeting, another conservative and religiously minded NSF member, Kevin McCann, conveyed even stronger reservations. McCann had been the president of Defiance College, a small college in Ohio associated with the Church of Christ. He had also been a key assistant to Eisenhower, both at Columbia University and then at the White House. And he had authored a glowing biography of Eisenhower supporting his successful bid for the presidency in 1952. Regarding NSF policy, McCann declared that "except for a few extremely limited areas," the social sciences represented "a source of trouble beyond anything released by Pandora." The agency, he added, "should not be fragmentized [*sic*] by a vain effort to comprehend the entire universe of human [re]search"—an argument reminiscent of the early postwar NSF debate, when some conservative critics pointed to the uncertain and undefined scope of the social sciences as one reason for excluding them in the proposed agency. In the midst of mounting discord, the NSB accepted McLaughlin's suggestion and told Hesburgh to study the matter further.[114]

After more work, Hesburgh's committee submitted a revised and scaled-back plan. The committee again called for an upgrade in organizational standing of the social sciences to office status and also greater proportional support for them in NSF's overall budget. As part of the push to extend social science funding, the committee had recently suggested that the agency should stop listing the specific fields of social research that were eligible for fellowship support, for this practice implicitly excluded other fields but without providing adequate justification. Nevertheless, the committee now backtracked and suggested continuing this practice. But most importantly, the committee rescinded its request to have "all of the various points of view and philosophies" represented on the external social science advisory committee.[115]

Though the NSB chair had warned Hesburgh that without further modifications even these modest ones might not pass, the board did, in fact, accept them. Accordingly, toward the end of 1958, the NSF created an Office of Social Sciences, now under Riecken's leadership. Early the following year, the NSB also declared, as noted before, that the social sciences would contribute to a more vigorous approach to social problems.

Nevertheless, those most responsible for advancing NSF's social science efforts found the repeated challenges tiresome, even demoralizing. In Riecken's words, "there seemed to be an endless need for justification, a continuous procession of questioners and critics." Time and time again, first Alpert and then Riecken had to defend their program, by explaining "the nature and purpose of social science," summoning "supportive statements from well known and prestigious figures," and asseverating "the scientific soundness, the uncontroversiality, and the intellectual promise of grants already made and of pending proposals." Bertha Rubinstein expressed the resulting frustration best when, after one "especially wearying reiteration" of the program's sound basis, she exclaimed: Sisyphus must have had the easier task.[116]

In short, the agency's internal policy-making dynamics ensured that the social sciences remained under careful scrutiny and their support strategically confined to the hard core. Father Hesburgh had tried to convince the agency's leaders to include a wider range of social science viewpoints on its external advisory board, based on his more general interest in supporting scholarship oriented to humanistic values and in accord with Eisenhower's reasons for appointing him to the NSB in the first place. Riecken noted that other voices in the "humanistic sector" also viewed the agency's "fledgling quantitative efforts with disdain, repudiating them as superficial, barren, wooden, and imitative of physical sciences—a form of intellectual social climbing."[117] However, a number of NSF's natural-science and conservatively minded leaders saw the effort to enlarge its social science program beyond the hard core as a threat to the agency's established priorities, the composition of its governing structure, and its reputation in Congress. Furthermore, hostile attitudes undermined the prospects for such reform, with the harshest words coming from President Eisenhower's close aide Kevin McCann. Under these conditions, any interest in supporting humanistic perspectives in the social sciences had little chance of gaining traction inside the agency.

"Official Negroes"

In the late 1950s and early 1960s, some social scientists also began to challenge the agency's limited engagement with their enterprise. After contributing so much to the development of NSF policies that cast the social sciences as junior partners in a unified and natural science–led enterprise, Harry Alpert would declare that the social sciences' second-class status had harmful consequences. From another direction, the political scientist Evron Kirkpatrick launched a campaign on behalf of his discipline by questioning NSF's determination to steer clear of potentially controversial areas of social research, especially politics. According to Kirkpatrick, this agency practice, which NSF leaders justified by appealing to a strict interpretation of basic science, amounted to discrimination against his discipline.

Shortly before Alpert left the agency in 1958, he published an article in the *Saturday Review*, a popular magazine, attacking the reigning scientific pecking order at the NSF and in the federal science establishment more widely. Though Alpert always expressed a firm commitment to the development of a scientifically progressive social science, his writings on Durkheim and other topics also recognized serious limitations in the unity-of-science viewpoint, at the levels of methodology, ontology, and social purpose. In his 1958 article, revealingly called "The Knowledge We Need Most," Alpert now wrote that the scientific hierarchy privileging the natural sciences had undermined a climate of professional respect and support for the social sciences. A sarcastic chart on NSF funding helped make his point. The chart depicts a man representing the social sciences who is located on a gently inclined slope and is just leaving the crawling stage, while the symbols representing the physical and life sciences soar on much steeper trajectories above him.[118]

Despite the humor, Alpert indicated that the social sciences' inferior position required urgent attention and correction. Toward this end, he called for steps to promote "a sense of mutual respect and understanding" between social and natural scientists and measures to erase the "invidious hierarchical distinctions among specialists of the various disciplines." The broader challenge involved establishing an "institutional environment" in which "all scientists" could work together in advancing the "understanding" of "nature, man, and society." Recently, the Soviet Sputniks had quickened American worries about a devastating attack from the skies. Referring to this awful possibility, Alpert proposed that "man's mastery of space" had to be "matched by man's mastery of himself. This is a fundamental condition of survival."[119]

After Alpert returned to the university world as a sociologist and high-level administrator at the University of Oregon, he often revisited these problems, arguing that the social sciences "suffered" because of their "minority group status" within American science. He pointed out that just like others who suffered from this handicap, social scientists were often the "last hired" and "first fired."[120] In addition, federal social research units still faced repeated criticisms, budgetary cuts, and closings, leading to what Alpert called "the 'on again, off again' pattern of government social science activities." Though no major anticommunist congressional investigations of social science patrons took place in the late 1950s, appropriation storms continued to threaten federally sponsored social research programs. A wide range of federally funded studies on topics such as child rearing, population growth, the diffusion of messages in a community, and mother love among lambs attracted legislative scrutiny.[121]

Unfortunately, given their political vulnerability, the social sciences had little incentive to try to stand on their own, Alpert added. After all, they "prospered best" when they resided under "broad umbrella classifications of the scientific disciplines such as the agricultural sciences, military sciences, medical sciences, and health sciences." "Close company" with the better-established natural sciences provided the social sciences with "protection and nourishment," which they rarely enjoyed when they stood "exposed," "naked and alone."[122]

Even after the Social Science Division's establishment in 1961, NSF's engagement with the "other sciences" remained limited in ways that confirmed Alpert's concerns and also prompted criticism from other social scientists who believed the agency treated them poorly. From the outset, the NSF had claimed a special interest in science as a whole.[123] As the agency grew in size and stature, it became known, in the words of American science policy historian Bruce Smith, as a "vital balance wheel among the various research support agencies." The agency sought to fill gaps in funding that arose because other federal agencies, commercial patrons, and philanthropic organizations had somewhat different interests.[124] As NSF's *1959 Annual Report* put it, the agency had the vital responsibility of "maintaining a comprehensive program in support of basic research."[125] Under these conditions, the established social science disciplines all had good reason to expect they would be welcomed into the new division. As indicated before, in the case of sociology, economics, anthropology, and the history, philosophy, and sociology of science, the division readily incorporated existing programs. But political science remained an outlier, a legacy of Alpert's early policy work and more general anxieties within the agency that identified politics, the subject matter of political science, as too hot to handle—sex, race, and religion were the other three forbidden topics during the agency's early years.

Not surprisingly, some disgruntled political scientists decided to challenge NSF funding policies. By the early 1960s Evron Kirkpatrick, the political scientist who had mentored Hubert Humphrey and also directed the State Department's Office of Intelligence Research, was serving as the American Political Science Association's (APSA) executive secretary. Finding his discipline's absence at the NSF unacceptable, Kirkpatrick launched a campaign for inclusion that persisted into the mid-1960s. Whereas Father Hesburgh had tried to broaden the agency's approach from the inside, political scientists had no standing there, and thus Kirkpatrick would have to try to apply pressure from the outside.

Before Kirkpatrick began his campaign, the rise of the behavioralist movement had already put the goal of gaining inclusion at the natural science–oriented agency within closer reach. As discussed in chapter 3, this movement gained a dominant though always contested presence within political science starting in the 1950s. Behavioralist scholarship reaffirmed America's special democratic heritage and exemplary modern character while also explaining how developing regions around the world could enjoy the fruits of America-friendly modernization schemes. Yet influential behavioralist figures also praised such work for leading the discipline toward the scientific high ground. Stimulated by major financial support from the SSRC and Ford Foundation, behavioralist political science scholars embraced quantitative methods of analysis, empirical hypothesis testing, and an allegedly value-neutral, nonideological investigative stance, purportedly marking a decisive departure from an older style of humanistic or liberal arts scholarship and thus presumably making such work attractive to the NSF. Indeed, in 1961 the NSF awarded its first grant of direct interest to political science for a proposal that glowed with behavioralist features. Put forth by the University of Chicago's Duncan MacRae, the proposed research aimed to "develop a rigorous statement of the relationship between popular and legislative votes . . . a mathematical statement of the process of representation."[126]

Notwithstanding MacRae's success in attracting funding, Kirkpatrick complained in a series of letters to NSF leaders written during the early to mid-1960s that political scientists remained at a distinct funding disadvantage when compared with other social scientists. Because the agency did not have a program for his discipline, in order to evaluate political science proposals the NSF had to rely on programs oriented toward other disciplines, which, as Kirkpatrick pointed out, almost surely resulted in unequal treatment of these proposals. Moreover, no NSF announcement or report had ever identified political science as eligible for support.[127]

Empirical evidence about the distribution of social science funding supported Kirkpatrick's gripe about unfair treatment. In 1961 the NSF awarded just 3 political science grants for a total of $75,000, while all the other social sciences combined received 140 grants totaling more than $3 million. The following year political science received only a single grant for $39,000, while the other social sciences received 194 grants for more than $7 million. Kirkpatrick acknowledged that only a small number of political scientists had submitted proposals. But he also reasoned that this fact reflected an appropriately skeptical attitude regarding the likelihood that their submissions would receive favorable consideration.[128]

In replying to Kirkpatrick's accusations, the NSF initially took the position that political science, by and large, lay outside the agency's interests because research in the discipline typically did not satisfy the agency's high standards of scientific merit and its basic research orientation. Cases such as MacRae's represented an important exception, though the agency also tried to reassure Kirkpatrick by claiming that proposals of such high quality could already be handled adequately by existing NSF programs.[129] But the Social Science Division director Henry Riecken also observed that it would be dishonest to include political science in a list of disciplines eligible for funding, because the agency had no intention of supporting research of an "applied, normative or policy-oriented" nature. Such research, Riecken added, accounted for most work in political science. Though Riecken provided no evidence to support this last claim, he tried to reassure Kirkpatrick by noting that NSF's Social Science Division had recently reached a consensus that political scientists should be subject to the same eligibility criteria as other social scientists.[130]

Beneath NSF's reluctant stance toward political science lay lingering concerns about the potential negative repercussions from funding studies on politically explosive topics. According to the minutes from a 1963 divisional meeting, political science was "not the only discipline that touches potentially sensitive areas," but it did "seem more centrally concerned with questions of public policy and partisan controversy." Accordingly, the agency should continue to proceed with caution. Although "scientific merit" should remain the agency's "basic criterion for judging a proposal," the agency also had good reason to take into consideration other practical issues as well. Specifically, the meeting minutes stated that "when the subject matter of the research becomes border-line and possible questions of sensitivity arise," the NSF should be reluctant to cross the border.[131]

Riecken elaborated on the practical concerns that still informed the behind-the-scenes interpretations of the agency's commitment to basic research. In a letter to the recently appointed NSF director Leland Haworth, who was a physicist just like his predecessor Waterman, Riecken wrote that the agency had "reasonable grounds for suspecting that Congress might not welcome government-supported research at least on American political questions." "Holding to a stringent definition of eligibility in terms of basic nature and scientific (rather than policy) orientation" thus seemed worthwhile, in order to minimize "the danger of a negative Congressional reaction." The Social Science Division had held to such a stringent definition "so far," and Riecken proposed that it should "continue" to do so.[132]

Because Congress held the purse strings, and because a group of legislators had already urged the NSF to establish a more catholic policy toward the social sciences, Kirkpatrick decided to ask for their help. In late 1963, after a few frustrating years of discreet letter writing to NSF personnel, he sent letters to various Congressmen, presenting them with financial facts about the lopsided distribution of NSF grants and pointing out the limits on its social science policies.[133] Early the following year, he continued his correspondence with director Haworth, now claiming that the agency's policies had resulted in "discriminatory hardship on political science as a discipline and a profession." In addition, he threatened to take further public action, this time by writing to all Congressmen and to all 11,000 APSA members.[134]

Though it would take more years of aggressive letter writing and the marshaling of external pressure from sympathetic congressmen and disgruntled political science scholars, Kirkpatrick's campaign for inclusion eventually achieved significant results. Most notably, NSF's Social Science Division finally established a political science program in 1966, thereby leaving behind one of the most visible manifestations of the agency's strategically narrow interpretation of its commitment to basic social science. No doubt, for scholars eager to pursue research in a scientistic vein, the establishment of NSF's political science program represented a major step forward. Yet Kirkpatrick's campaign never challenged the agency's policy of limiting funding to studies at the hard-core end of the social research continuum, as defined in agency documents throughout this entire period as research that met the allegedly universal scientific criteria of objectivity, verifiability, and generalizability.

For scholars who valued various types of inquiry that did not satisfy such criteria, then, the agency thus remained a major source of discrimination. As reported in the news column of the *American Behavioral Scientist*, despite the agency's official reassurances about fair treatment, social scientists continued to "resemble the 'official Negroes' of NSF grants policy." Any social scientist invited to an "NSF cocktail party" would need to "wear the proper scientific garb," would have to "avoid political science and all controversial topics," and would need to be "properly respectful and grateful."[135] While Harry Alpert never spoke in such harsh terms, he too had concluded that the social sciences paid a high price for the protection provided by their close proximity to the natural sciences.

Conclusion

In the story of social scientists' midcentury quest for scientific status, political acceptability, and funding, NSF's development and its strategy of supporting hard-core social research played an important role. Crafted by Harry Alpert, this strategy initially consisted of modest support for convergent research, meaning studies that had significant overlap with the physical or biological sciences. As noted later by Henry Riecken, the agency's circumscribed approach to the social sciences led some scholars to believe the agency was "merely paying lip-service to the whole subject." However, its social science efforts during the late 1950s and early 1960s expanded in significant respects, as the agency dropped the convergent criteria, increased funding levels, and gave the social sciences stronger organizational status, culminating in the creation of a Social Science Division in 1961. These developments, together with the agency's growing reputation as a patron of first-rate science and especially university science, encouraged "more established, prominent and reputable social scientists to apply" for support.[136] Thus, in contrast to the Ford Foundation's BSP, whose emphasis on supporting scholarly work encountered serious difficulties that led to the program's tilt in an applied direction and then to its premature closure, NSF's social science program grew in importance, providing a welcome measure of financial support and scientific legitimacy to scholars, projects, methodologies, and fields of study that met the agency's hard-core criteria. Given the predominant emphasis on action-oriented and applied studies within defense science agencies as well, NSF's consistent dedication to funding basic social research also stood out.

Though the NSF and its high-profile scholarly supporters such as Pendleton Herring and Dael Wolfle emphasized that the agency promoted the social sciences in accordance with the most rigorous scientific principles, we have also seen that its hard-core emphasis reflected a variety of political, institutional, and practical constraints. The legacy of the postwar NSF debate over whether to include the social sciences, the new agency's natural-science orientation and its cautious leadership, paltry budgets during its early years, McCarthyite political scrutiny, and the presence of skeptical and sometimes downright critical perspectives within the agency's governing board: all these factors encouraged the agency to place strict limits on its social science activities. Though the agency's recognition of basic science principles, such as the oft-mentioned trio of objectivity, verifiability, and generalizability, also appeared in more scholarly oriented discussions, the agency relied on such criteria for political and institutional purposes that went beyond considerations of intellectual quality or scientific merit alone.

Indeed, the NSF really did not aim to advance the social sciences on a broad front, despite the claim by director Waterman and agency documents that stated it supported basic research in all the sciences in a comprehensive manner. In developing a viable funding policy, Alpert, Riecken, the agency's external social science advisers such as Herring, and NSF's natural science–oriented leaders tried to protect the agency from criticism by adhering to a model of inquiry that systematically excluded a variety of topics and methodological approaches deemed too controversial from a political or scientific viewpoint. Humanistic or philosophical perspectives thus remained ineligible for support. So did policy research, normative studies, and otherwise value-laden inquiry. For a number of years, the agency considered research on sensitive topics like race, sex, religion, and politics to be off-limits as well, regardless of the researcher's scientific qualifications or the study's particular methods and aims.

This chapter has also explored criticisms of the NSF that reflected broader challenges to the scientistic project in Cold War America. Even though the agency never ventured beyond the hard core, cautious voices and more strident conservatively minded critics from the natural-science and political communities suggested that the agency's approach should be more restrictive or at the very least should not be allowed to expand any further. In private correspondence from 1960, Father Hesburgh claimed the NSB "almost threw out the baby with the bath."[137] Though I have found no evidence to confirm that the board ever seriously considered axing the social science program altogether, the thought could not have been far from the minds of those such as Kevin McCann.

Two other lines of criticism suggested just the opposite; namely, the agency should expand its social science support and relax its hard-core and scientistic emphasis. After having been appointed to the NSB because President Eisenhower wanted the distinguished Catholic theologian's humanistic perspective at the agency, Father Hesburgh used his authority as chairman of a special NSF committee to propose that the agency's external social science advisory board should include all of the various points of view and philosophies represented. Alpert put forth a somewhat similar challenge, though from a secular perspective more characteristic of liberal critics of scientism. While NSF's first social science policy architect had always recognized that the social and natural sciences had much in common, he also identified some important epistemological and methodological differences between them that arose from the need for interpretation in social inquiry. In his scholarly writings where he was not representing the NSF, Alpert stated that in order to promote liberal democratic values, the social sciences could not be constrained by firm distinctions between scientific knowledge and value-laden social inquiry, social reform, and social criticism. Toward the very end of his tenure at the NSF and in subsequent years, Alpert also claimed that natural-science envy and the scientific pecking order undermined the healthy development of the social sciences.

While the criticisms from Hesburgh and Alpert had little impact on the agency, Evron Kirkpatrick's campaign did prove effective in challenging the agency's cautious stance regarding political science research, though the resulting reforms left NSF's scientistic policy framework firmly intact. Before Kirkpatrick's campaign, the NSF had supported only the occasional political science project marked by features associated with the behavioralist revolution. During the early 1960s Riecken argued that in order to avoid unwanted political scrutiny, the NSF should continue to avoid many areas of political science research because they dealt with controversial matters, a position reaffirmed by the NSF leadership. As a consequence, though all the social science disciplines supposedly had equal status at the agency, some disciplines remained more equal than others in practice, with political science at the bottom of the pack. Even after Kirkpatrick's campaign succeeded in getting political science its own program within the Social Science Division, NSF's support for political science remained well below its support for other disciplines. Moreover, the agency continued to channel support to research with a behavioralist orientation.

CONCLUSION

In this book I have argued that the military, the Ford Foundation, and the NSF became major players in a significantly transformed and largely new extra-university funding system for the social sciences in Cold War America. The development of these important patrons presented social scientists and other interested parties, including natural scientists and politicians, with critical opportunities to work out the nature and uses of social science research and expertise in the nuclear age. In addressing these basic issues, patrons together with their scholarly collaborators became strong proponents of a scientistic strategy for advancing the social sciences and a closely associated social engineering viewpoint, which explained how advances in knowledge production would contribute to social progress and the national welfare. In turn, these commitments became central in the evolution of the politics–patronage–social science nexus. The first part of this conclusion highlights the main lines of my analysis concerning the development and importance of patron policies and programs that supported these commitments. In the second part I review five main challenges to the Cold War patronage system discussed in this study, and then indicate briefly how those challenges informed more substantial efforts to rethink and reform the politics–patronage–social science nexus starting in the mid-1960s.

Because scientism and social engineering commitments played a vital role in the earlier period of professionalization in American social science, one might be tempted to see a smooth development of those commitments from the prewar to the wartime and then to the postwar years. However, I have explained why we should resist this interpretation. To be sure, prior scientistic and social engineering commitments provided an appealing legacy to their advocates after World War Two. When scholars from the social sciences discussed the basis for future progress in their fields, they sometimes invoked prewar precedents, and they often pointed to their wartime work as a model. Nevertheless, the power of those commitments during the early Cold War decades should not be regarded as simply an outgrowth or direct extension of earlier trends, partly because the interwar era included too many dissenting voices, but also because early postwar discussions about the social sciences and their position within the rapidly changing political and scientific landscapes revealed great uncertainties. Under these conditions, social scientists, natural scientists, members of congress, and various other parties engaged in contentious debates about the scientific identity, practical value, and national contributions of the social sciences. Those debates along with the development of specific funding policies and programs became critical in establishing the widespread acceptance of those scientistic and social engineering commitments in the new patronage system, which, in turn, encouraged their prominence throughout the social and psychological sciences during the first two Cold War decades.

In considering how these two ideas acquired such prominence, I argued that the postwar NSF debate played a key role, because it helped to establish the problematic relationship of the social sciences to major developments in partisan politics and national science policy. During the New Deal, the social science–led National Resources Planning Board had been in a position to pass negative judgment on an ambitious science policy proposal from natural scientists. But in the nuclear age natural scientists and especially physical scientists, including an influential conservative contingent who supported Vannevar Bush's science policy plans, held much more powerful positions than social scientists in the nation's science policy affairs. Moreover, many figures from the natural-science, engineering, and medical communities who supported Bush's plans challenged the scientific credentials of the social sciences and attacked their involvement with liberal or leftist politics. The NSF debate also helped to forge an alliance between those scientists and conservative politicians who attacked the social sciences for some similar reasons stemming from their broad antipathy toward the New Deal and its legacy in various wartime and postwar initiatives and programs. Meanwhile, though the social sciences had some allies, mainly liberal and left-liberal scientists and politicians, their support proved to be rather ineffective. As a result, the alliance of critics managed to place the social sciences on the sidelines of the ongoing NSF debate rather quickly.

In addition, during this episode the Social Science Research Council (SSRC) and its prominent social science representatives embraced a scientistic strategy for gaining inclusion in the proposed agency. In their congressional testimony, the SSRC group, led by Wesley Mitchell, insisted on the objective, nonideological, and value-neutral character of modern social science inquiry and emphasized its distinctiveness from programs of social reform, socialism, normative inquiry, and humanistic studies. In various writings, including his SSRC-commissioned essay, Talcott Parsons also advocated a scientistic stance and took care to dismiss conservative criticisms that portrayed the social sciences as hotbeds of radicalism led by long-haired men and short-haired women. Equally important, the SSRC managed to keep the views of social science critics of scientism out of sight. These critics included Louis Wirth, who tried to redirect SSRC's efforts away from what he and other well-known liberal critics, such as Robert Lynd and Gunnar Myrdal, took to be a rather narrow, overly technical, and thus woefully impoverished vision of social science and its social responsibilities.

Of course, the NSF debate by itself did not directly determine much about the actual course of postwar funding. Furthermore, in the late 1940s and early 1950s the patrons considered in this study all faced formidable challenges in figuring out what type of scientific knowledge and practical value the social sciences had to offer. Though social scientists had participated in a variety of wartime activities, by the early postwar years many social science–oriented programs and projects within the federal government had either come to an end or else suffered sharp reductions in scope and size. When the Human Resources Committee of the military's Research and Development Board (RDB) held its first meeting in 1947, Vannevar Bush, speaking as RDB's first chairman, asserted that the relevant field of interest, namely, the social sciences, contained some kooks. Thus figuring out how to proceed in this field presented unusual difficulties. Drawing on previous debates during the 1930s, extensive controversy about the course of philanthropic funding for the social sciences arose in the early postwar years as well. After World War Two, during a prolonged process of internal review at the Rockefeller Foundation, influential figures including Raymond Fosdick and Chester Barnard criticized the scientistic orientation as misguided, ineffective, and also injurious because it undermined the ability of social scientists to address many urgent issues including a worldwide crisis in morals. Following the divisive NSF controversy, whether the new agency would support the social sciences and how it might do so also remained unclear.

Nevertheless, the patrons examined in this study all decided to pursue a strategy for funding the social sciences in a manner strongly foreshadowed by the postwar NSF debate. Though the level of personal commitment on the part of individuals who participated in the elaboration and implementation of this strategy varied, I have argued that the subordinate position of the social sciences in the recently transformed and greatly enlarged federal science establishment made that strategy widely attractive. As the Cold War funding system took more definite shape, social scientists and their advocates regularly found themselves on the defensive, constantly struggling to establish their scientific credentials in a funding system in which the major participants tended to take the natural sciences as the gold standard. Moreover, wherever they enjoyed substantial power, the nation's postwar natural-science elite showed little interest in allowing social scientists to obtain much influence, as we have seen in the cases of the defense science establishment and the NSF as well as in the National Academy of Sciences and President's Science Advisory Committee. Although at the Ford Foundation natural scientists lacked the dominant presence they had within the federal science arena, the physicist Karl Compton, an influential member of Bush's group and a strong critic of the social sciences, had a seat on the board as the modern Ford Foundation took shape in the late 1940s. In addition, Henry Ford II followed Compton's recommendation to hire H. Rowan Gaither to lead the study team responsible for charting the foundation's new direction. Gaither, who had close ties to many physical scientists of the Manhattan Project generation and also to postwar military science programs through his position as chairman of the new RAND Corporation, held greater sway at Ford through his multiple roles: by directing the team that produced the landmark 1949 *Study Report*, by helping to set up the new Behavioral Sciences Program (BSP) and recruiting its first and only director, Bernard Berelson, and by serving as Ford's president from 1953 to 1956.

Equally important, persistent criticisms from various conservative quarters together with anticommunist politics encouraged social scientists and their patrons to distance their activities from liberal ideology, left-liberal politics, and more radical leftist programs. The involvement of social scientists in a wide range of New Deal, wartime, and postwar projects and agencies, from the National Resources Planning Board to the Office of War Information, the Department of Agriculture, and the Full Employment Act, all provoked spirited attacks from conservative interests in the political, scientific, business, and intellectual communities. The postwar Red Scare and McCarthy Era supported harsh scrutiny of many social scientists and especially those associated with progressive positions on a wide range of subjects, including race relations, public housing, labor unions, economic regulations, social security, codes of sexual conduct, civil rights for those suspected of subversion, and the international control of atomic energy. In the early to mid-1950s anticommunist investigations led by conservative legislators from both major political parties also created serious headaches for social science patrons, especially the large private foundations including Ford. From time to time the military faced pointed scrutiny of its social science activities, as indicated by the bitter complaints from RDB's Human Resources Committee about legislators not understanding the social sciences and as seen more directly in the episodes that prompted closure of the air force's Human Resources Research Institute (HRRI) and provoked alarm about the RAND-sponsored study on unconditional surrender. At the NSF social science advocates also faced pressures from a mix of conservative positions in the political and scientific communities. All the while, social scientists and their patrons confronted a variety of official mechanisms designed to expose and undermine subversive influences, including background checks, security clearances, and lists of suspect individuals and organizations generated by the FBI and the U.S. attorney general.

Those developments in American science, federal science policy, partisan politics, and anticommunism inspired patrons and social scientists who worked closely with them to engage in extensive boundary work. The governing boards, top staff members, program managers, and scholarly consultants for the patrons considered in this study all proposed that social scientists should adopt strictly scientific attitudes, investigative tools, and research goals in a manner that would establish their rightful place within a unified scientific enterprise and would establish a clear boundary between scientific social inquiry on the one hand and politics, ideology, reform, philosophy, and mere social opinion on the other. These basic points appeared in a thousand different places, in testimony before congressional committees, in patron annual reports, in other publications from patrons and their social science representatives, in internal documents and personal correspondence concerning patron policies and programs, and in national statements about the social and behavioral sciences. This concerted effort also shaped the search for terminological innovations: military documents often referred to "human resources research" and "the behavioral sciences"; the latter term became most closely associated with the Ford Foundation's BSP; and the NSF declared that it supported only work at the "hard-core" end of the social research continuum.

Scientistic and social engineering commitments gained prominence also because the social scientists who helped to shape, direct, and promote new funding programs shared a great deal in common. Through their academic careers and other professional experiences, these scholars had intimate knowledge of the changing politics of American science during the wartime and Cold War years. And they often knew one another personally through overlapping professional networks. Thus numerous scholars who worked with the military had ties to foundation programs as well, including Bernard Berelson, Charles Dollard, Pendleton Herring, Donald Marquis, Hans Speier, and Donald Young. Individuals who shaped NSF's social science efforts had significant ties to the military or the private foundations and sometimes to both, including, again, Dollard and Herring but also Harry Alpert, Chester Barnard, Henry Riecken, and John Wilson. I am not suggesting they all had identical views about the social sciences, never mind what they may have thought about specific fields of research or particular projects. Nevertheless, we have seen they had a lot in common as they tried to craft an approach to the social sciences that patrons could find attractive and that many other leading scholars from the social and behavioral sciences could welcome as supportive of their professional ambitions. Furthermore, even when some of them had reservations about the unity-of-science viewpoint and the associated hierarchical relationship that relegated the social sciences to an inferior position vis-à-vis the natural sciences, they still recommended funding policies aligned with a scientistic viewpoint owing to strategic considerations, as seen most clearly in the cases of Barnard and Alpert at the NSF.

Furthermore, patrons and their program managers stuck closely to the scientistic strategy in practice, through the selective allocation of funds to various levels of the social science enterprise. Of course, it is always possible for individuals charged with running funding programs to interpret policy guidelines and grant-making criteria loosely, in order to pursue promising opportunities otherwise prohibited by too strict an interpretation. Likewise, by "working the system" scholars can get funding for projects that would otherwise be poor candidates for support had their applications described these projects more accurately. But scientistic and social engineering commitments had a profound impact because patrons and scholars did, in fact, implement them in practice. We have seen substantial evidence of this reasonably tight match linking the boundary work, rhetoric, and policies of patrons to what they actually funded at many levels, including their support for major social science institutions, organizations, and centers such as the SSRC, the Center for Advanced Study in the Behavioral Sciences, the Population Council, Harvard's Department of Social Relations and Russian Research Center, MIT's Center for International Studies, the Office of Naval Research (ONR), the air force's HRRI, RAND, and the army's Special Operations Research Office (SORO); for many important fields of inquiry including development studies, behavioralism in political science, juvenile delinquency research, communications studies, operations research, systems analysis, game theory, deterrence theory, counterinsurgency studies, "convergent research" as defined at the NSF, methodological work focused on mathematical and other forms of quantitative analysis including survey research, and also the history, philosophy, and sociology of science; and for a long list of individual scholars whose reputations as leaders in their respective disciplines and interdisciplinary fields of study benefited significantly from their ability to cultivate patron support.

We have also observed some salient differences among major patrons and their engagements with the social sciences. A great deal of military funding and research focused on the sciences of choice, which aimed to place military strategy and operations on a rational basis. Other military-funded work concerned the sciences of control or, as Charles Bray put it, a technology of human behavior, which promised to facilitate effective manipulation of human behavior and thought, human-machine interactions, and social systems marked by their national security importance. Particular military units had their own mandates that shaped their involvement with the social sciences in distinctive ways as well. For example, in the 1950s the air force–funded RAND Corporation became deeply involved with developing strategies for fighting or deterring nuclear war that depended on mathematical and economic tools of analysis, though RAND also funded many other types of work. Meanwhile, the army's

SORO concentrated on the challenges of fighting ground wars and became known for its work on counterinsurgency studies that incorporated a mixture of behavioral science approaches. At the Ford Foundation, BSP's initial policy emphasized the need to develop the basic resources in the behavioral sciences, whose advance would then inform the more practically oriented work pursued through Ford's other programs. But those other programs themselves also supported an enormous amount of research, for example in the areas of modernization studies, economic development, urban decay, and juvenile delinquency. The NSF concentrated on funding basic inquiry and especially hard-core social research, leaving the responsibility for funding practically oriented endeavors to other patrons, including the military and the large private foundations. For a number of years the NSF also steered clear of research on certain sensitive topics, including sex, race, religion, and politics, at a time when common wisdom suggested that private patrons could support research on such sensitive topics more effectively.

Yet, it is worth emphasizing again that beneath some noteworthy differences in what patrons funded lay a common and widespread strategy about how to promote the development of the social sciences in order to realize their scholarly and practical potentials. Thus those differences among patrons did not indicate their support for alternative understandings of the social sciences. Quite the contrary, in fact.

Indeed, within the Cold War funding system explicit alternatives to scientistic understandings of the social, behavioral, and psychological sciences attracted little positive attention or material support. Moreover, the marginalization of alternative perspectives occurred not as an unintended result or accidental byproduct but, instead, as a direct result of the decisions taken by those directly responsible for funding policies and programs. So, as we have seen, programmatic statements, assessments, and recommendations concerning military research regularly noted a sharp difference between scientific social research and humanistic types of inquiry and then suggested the military should focus on the former. Despite some early discussion at the Ford Foundation about building ties between the humanities and the behavioral sciences, BSP's leaders never pursued this possibility seriously. At the NSF, those individuals who wanted the agency to welcome viewpoints beyond the hard core got nowhere, as indicated most vividly by the episode involving Father Hesburgh.

The lines of analysis reviewed above have already indicated why these patrons deserve significant attention in our histories of the social sciences in Cold War America. This study has also offered some additional reasons why the Cold War patronage system deserves attention in our narratives about the politics of knowledge in American society and government during the first two Cold War decades. For one, advocates of scientistic and social engineering approaches argued that these approaches had a crucial role in advancing the social sciences as a valuable national resource. Parsons's SSRC-commissioned essay even featured this point in its title, "Social Science: A Basic National Resource." Elsewhere, scholars and patrons claimed that the social and psychological sciences had vital roles to play on the Cold War battlefield, for example by contributing crucial expertise needed to win the struggle for hearts and minds and by ensuring that the communists did not obtain a crucial advantage in the race to develop ever more effective techniques for controlling mind and behavior. In a similar vein, but with greater attention to domestic issues, patrons and their scholarly collaborators proposed that these sciences would strengthen the nation's ability to address serious problems such as mental illness, juvenile delinquency, ideological fanaticism, and political radicalism.

Sometimes the Cold War served as an important reference point in these discussions. At a time when American scholars, politicians, and media commentators commonly asserted that the communist enemy and its scholars remained wedded to a misguided and dangerous Marxist ideology, American social scientists and their patrons argued that they embraced a strictly scientific and therefore vastly superior approach. Parsons, Herring, and Berelson all argued that the general trend among American scholars wisely rejected an earlier interest in grandiose, ideological schemes for social transformation grounded in all-encompassing theories of historical change. Focusing on more limited and hence tractable research studies, these scholars were, instead, constructing a truly scientific enterprise that promised to produce trustworthy results of great national value.

The same understanding also informed the claim that American social science had great value as an export to other regions around the world, another point that scholars and their patrons often made with reference to the Cold War struggle. According to this general idea, American social scientists had developed a powerful, nonideological investigative model that other countries, whether in war-torn Europe or in the developing regions of the world from Latin America to Southeast Asia, would be wise to embrace. To be sure, scholars did not need to have close relations with patrons to make such claims. However, these claims gained traction partly because of explicit encouragement from patrons and the scholars they supported, especially those who worked in international contexts, including the field of development studies, where great effort focused on elaborating America-friendly modernization programs for the third world, as the antidote to socialist planning and Marxist doctrine.

Furthermore, such claims about the social sciences' scientific foundations, practical value, national contributions, and relevance to international affairs acquired growing importance in the nation's affairs. During the early 1960s President Kennedy became well known for his eagerness to work with scholars who could bridge the worlds of academia and politics comfortably. Though President Johnson did not feel so comfortable in the company of action intellectuals from elite universities, he came to depend even more heavily than Kennedy on advisory groups and individuals with impressive scholarly credentials, including many social scientists who had a hand in designing major domestic initiatives, including the War on Poverty and other Great Society programs, and in crafting equally noteworthy foreign policy ventures, including foreign aid programs for developing countries and the war in Vietnam.

*****

Yet we have also seen that the Cold War funding system and individual patrons faced various criticisms and obstacles, which I have discussed in the form of five challenges. Below, after first reviewing these challenges, I will indicate how they informed the growth of more substantial opposition to the politics–patronage–social science nexus starting in the 1960s. I also discuss a few ways in which subsequent efforts to reshape the extra-university funding landscape helped to make this latter period distinctive in the evolution of the American social science enterprise.

In their efforts to defend themselves against conservative criticisms from the natural-science and political communities, patrons and their scholarly collaborators often claimed that social and behavioral science inquiry could and should be value neutral and apolitical, just as natural-science inquiry seemed to be. Yet, as we have noted, certain conservative criticisms, fed by a number of tributaries, characterized scientism itself as misguided and terribly harmful, thus presenting one major challenge. In the first decade after World War Two, this viewpoint received significant attention in various settings, including in the internal discussions and assessments of social science programs at the Rockefeller Foundation and in a series of McCarthyite investigations. The Reece Committee investigation provides a particularly rich example, as Representative Reece, his loyal research staff, and some critics of scientism from the social science community rejected the claim of a presumed unity between the social and natural sciences, because of alleged fundamental differences in their respective subject matters, methods of inquiry, knowledge claims, and practical uses. Furthermore, the Reece Committee's case against scientism had a sharp political and moral edge, for its proponents asserted that the social sciences' pursuit of strictly scientific knowledge, together with the resulting marginalization of humanistic modes of inquiry, had promoted various ills such as godlessness, excessive expansion of governmental power, socialist regulation of the economy, liberal social engineering, sexual immorality, and moral relativism. During the late 1950s and early 1960s, social scientists enjoyed a respite from anticommunist investigations. Nevertheless, critiques of scientism remained influential in certain conservative political, business, and intellectual circles.

Meanwhile, some liberal scholars raised a second challenge by questioning the unity-of-science perspective and its prominence in patron-funded scholarship. In the postwar years, left liberals including Gunnar Myrdal, Robert Lynd, and Louis Wirth argued that social scientists should pursue their research in ways that explicitly advanced values and ideals they associated with the nation's unfulfilled democratic mission and the promise of equality and opportunity for all citizens, regardless of race, ethnicity, or creed. From this standpoint, social inquiry could be scientific without imitating the natural sciences and without leaving aside direct engagement with humanistic values, democratic ideals, and the practical means for realizing them. Thus these scholars rejected the sort of objectivity that rested on a value-neutral and allegedly disinterested scholarly stance. They further argued that when scholars tried to excise values from social inquiry, their biases and hopes regarding human nature and the social order often crept in through the back door, and sometimes with alarming antidemocratic implications, as Lynd found in the case of the *American Soldier* studies carried out during the Second World War. Though not well supported in the political and academic arenas during the McCarthy Era, this critique received increasing attention among intellectuals by the early 1960s and found an especially receptive audience in the New Left. The sociologist Irving Louis Horowitz, a prominent social science advocate during the early years of the New Left, became a major critic of military-funded social research that, in his view, hid its biases with a veneer of scientific objectivity, value neutrality, and quantitative precision. Horowitz roasted scholars associated with RAND and other military–funded research centers for producing work that seemed hopelessly unrealistic in its suppositions about human nature and international affairs and dangerously militaristic in its implications for American Cold War strategy.

Some scholars, though they seemed quite comfortable with the scientistic strategy, also recognized that the power of patrons presented certain threats to academic autonomy and scholarly integrity, thus posing a third challenge. As noted by John Darley, the psychologist who chaired ONR's Advisory Panel on Human Relations and Morale, government funding could lead to unhealthy government planning and control of research. When discussing this issue, social scientists often indicated that such threats could be managed effectively, owing to a number of considerations: because all patrons relied on social scientists as consultants and program leaders; because patrons provided some significant support for basic social research; because in many cases funding for practically oriented research did not preclude scholars from making valuable academic contributions; and because a measure of pluralism in the Cold War patronage system promised to mitigate the impact of any individual patron that disrespected the intellectual, institutional, and social conditions that contributed to excellence in academic scholarship. Of course social scientists who appreciated the support provided by patrons also had practical reasons for not speaking too critically about the threats they posed. Remember that after Wesley Mitchell's SSRC committee had expressed concerns about the potential dangers of dramatic increases in federal funding and control of social science research during the mid-1940s, subsequent SSRC discussions led to the conclusion that expressing reservations about these dangers in public might hurt the social sciences' prospects of obtaining public funding, which, in turn, would be disastrous to their future well-being. Yet such threats remained ever present. As the years passed, they also became more problematic, especially in the case of the expanding military–social science partnership. Military support for research with strategic relevance had an especially high likelihood of attracting critical scrutiny whenever the relevant strategies became controversial. In addition, classified research funded by defense and other national security agencies, such as the CIA, posed an especially strong threat to the freedom of scholarly discussion, criticism, and evaluation of scientific research.

Fourth, social and behavioral scientists rarely enjoyed more than a marginal status within the federal science establishment throughout the first two Cold War decades. The distribution of federal funding across the sciences and the numbers of representatives from the different sciences within particular science agencies and science advisory bodies made the reigning scientific hierarchy perfectly clear. When social scientists tried to make a strong case for significant expansion in their funding and representation in these settings, the ensuing discussions typically ended in frustration. After the demise of the Ford Foundation's BSP, the young natural science–oriented NSF acquired greater national importance as a patron for basic social research. But compared with the agency's other science programs, its social science efforts were very small, and their expansion over time remained carefully restricted, leading one unhappy scholar—probably a political scientist—to complain that social scientists remained the agency's "official Negroes." On another occasion, NSF's longtime social science assistant Bertha Rubinstein remarked, after one trying meeting with the agency's natural science–oriented governing board, that Sisyphus had had the easier task. In a less colorful but similar manner, the military psychologist Charles Bray observed in the early 1960s that social and psychological scientists continued to struggle with the problem of having to please natural scientists and engineers who dominated the defense science establishment and who were often unsupportive of social science work.

The fifth and final challenge arose from the blurring of the boundary that supposedly separated scientific social research from a host of other activities including value-laden inquiry, social reform, ideology, and politics. Of the patrons considered in this study, only the NSF maintained a consistent reluctance to support research that engaged directly with questions about how to live and what should be done. Indeed, this reluctance frustrated critics who suggested the agency's social science efforts should deal with a wider range of viewpoints and approaches. And one could dig deeper into particular NSF-funded projects to see whether, despite initial appearances, they reflected value judgments or political objectives. Still, the agency's cautious leadership and its social science advisers insisted that the agency should steer clear of value-laden, prescriptive, and action-oriented work. But various lines of research funded by the defense science establishment and the Ford Foundation took a different direction that reflected a more direct interest in strategy, policy, and action. Thus the Defense Department, the individual military branches, and the Ford Foundation supported a great deal of research that aimed to define, promote, and secure goals associated with the Cold War struggle and other efforts to advance American interests and ideals around the world. As seen in such high-profile examples as development studies and counterinsurgency research, social science researchers and patrons often emphasized that the type of knowledge sought and its anticipated uses had direct relevance to the great political and ideological conflicts of those years. Ford also supported various lines of social and behavioral research that had explicit domestic policy implications. Moreover, Ford's leadership strongly favored work that promised practical payoffs, which caused deep disappointments with the BSP and ultimately helped to bring about its demise. None of these observations implies that researchers and their patrons had abandoned the effort to advance the social sciences as a scholarly enterprise. However, in many cases, patron-funded work did move away from a value-neutral and disinterested stance in favor of a more engaged brand of scholarship.

During the turbulent 1960s, those five challenges provided crucial background for mounting criticisms of the politics–patronage–social science nexus. As the 1960s opened, social scientists' involvement in major new domestic and foreign policy initiatives made them much more visible and influential in American public life. But these developments also inspired widespread criticisms of their work and policy contributions as those same initiatives came under attack from various viewpoints. According to their strongest critics, growing social science influence in domestic and foreign policy affairs had actually made the nation's problems worse than before. As mounting racial tensions produced race riots in American cities from coast to coast, as significant pockets of poverty persisted in an otherwise affluent society, as the Vietnam War raged on despite massive American military commitments and official declarations of imminent victory, as federal expenditures on war and social programs contributed to mounting inflation, and as development programs for the third world strengthened the hands of brutal dictators, critics claimed social scientists and their patrons had become part of the problem rather than the solution.

The most visible critics of social scientists' policy-oriented work during the mid- to late 1960s commonly leaned to the left, as opposed to the greater prominence of right-wing critics during earlier Cold War years. A collection of worries presented by the growing chorus of concerned liberals and more radical voices suggested that connections to powerful patrons, especially the military and intelligence agencies but also the large private foundations, had turned social scientists into tools of the Establishment, dedicated to maintaining an unjust status quo at home and abroad. At mid-decade national media coverage and a congressional investigation of Project Camelot, an army-sponsored research project planned by SORO, stimulated scrutiny along these lines. Worried voices from the social sciences, politics, and mass media, as well as some foreign intellectuals and governments, charged that such work reflected and contributed to American arrogance, militarism, and neocolonialism. Thus, rather than supporting social science of a nonideological and apolitical character, the military seemed to be funding work bathed in a potent pro-American, imperialist, and antirevolutionary ideology. A few years after his scathing attack on RAND's civilian militarists, Horowitz criticized the military–social science partnership in numerous publications including a well-known edited book, *The Rise and Fall of Project Camelot*, whose essays from American and non-American authors provided critical accounts of this project, counterinsurgency research, and the military–social science partnership more generally.[1]

While the controversy over military funding gathered force, a predominantly liberal group of social scientists and politicians tried to rectify the second-class status of the social sciences within the federal science establishment. Typically their efforts combined confidence in the ability of social science to help government solve national problems with concern about the reigning natural science–dominated scientific hierarchy. One of the most ambitious efforts, though it ultimately failed, involved a legislative proposal to create a National Social Science Foundation, put forth by the Oklahoma Democratic senator Fred Harris. He and many advocates of his proposal, including some prominent social scientists working in government and in the academy, argued that the nation should provide the social sciences with much greater support, that social scientists should have much stronger representation within the federal science system, and that the proposed agency should not limit its support to research with a scientistic outlook. Harris developed this last point with specific reference to the restrictions at the NSF, where a scientistic outlook remained firmly in place.[2]

A powerful resurgence in American conservative thought and politics inspired additional charges against the allegedly objective, nonideological character of the social sciences and their policy influence. One prevalent line of criticism found liberal and radical social scientists guilty of promulgating their agendas under the guise of scientific research. Whereas critical liberals and more radical voices charged that large parts of American social science bore the ideological imprint of the Establishment, conservative voices typically suggested the very opposite, namely, that misguided left-wing scholars and their equally misguided patrons posed a threat to the nation's venerable institutions and admirable values. Conservatives of various stripes blamed social scientists for contributing to a variety of problems, such as encouraging racial strife; undermining the stability of the family; promoting expensive, ineffective, and harmful government polices; destroying American capitalism; and weakening the nation's resolve to defeat the communist menace.[3] In these discussions, one can hardly miss the echoes of earlier conservative, anti–New Deal, and anticommunist attacks on social scientists and their patrons.

Following the mid-1960s, social science patronage evolved in response to the challenges noted above in ways that make this later period significantly different, though not to say discontinuous, from the previous one. By the late 1960s a number of universities, in the name of demilitarizing academic social science and protecting their reputations, severed their formal ties with military-funded social research institutes. For example, American University refused to provide SORO with a home any longer.[4] Another reconfiguration in the funding landscape became visible in the 1970s, when a wave of new or newly enlarged foundations and think tanks with an explicitly partisan character appeared. The most influential of these organizations championed viewpoints associated with the libertarian, free-market, neoconservative, and traditionalist strands of American conservatism.[5] Their partisan commitments represented a striking departure from the earlier tradition developed by the large philanthropic foundations, including Carnegie, Rockefeller and Ford, as well as by private research organizations, such as the Brookings Institution and the National Bureau of Economic Research, which had all claimed a commitment to objective and nonpartisan social inquiry.

A third trend involved not a rethinking of the scientistic outlook but a reaffirmation of its value. At the NSF, a noteworthy case, social scientists during the 1970s and 1980s found themselves under sharp attack yet again from skeptics, especially conservatives who questioned the scientific status of their work, its practical value, and its political implications, much as their predecessors in the 1940s and 1950s had done. Under these conditions, social scientists, who continued to struggle for recognition and funding in an agency dominated by natural-science interests, once again embraced the unity-of-science viewpoint and argued that the agency's promotion of objective, apolitical social inquiry of the highest scientific caliber constituted a national resource.[6]

More recently, the role of social scientists in the War on Terror has attracted critical scholarly, media, and political attention. Following officers' complaints of poor local intelligence in Iraq, the Defense Department established the Human Terrain System (HTS). Initially, this program recruited scholars from anthropology, political science, sociology, geography, and linguistics to participate in teams assigned to U.S. combat brigades in Iraq and Afghanistan. In a manner similar to that of the earlier liberal and radical critics during the 1960s, some scholars have characterized HTS as the latest effort to "weaponize" the social sciences. In the words of anthropologist Hugh Gusterson, "the Pentagon seems to have decided that anthropology is to the war on terror what physics was to the Cold War."[7] In 2007 the American Anthropological Association's executive board issued a harsh assessment with the following concluding points: "the HTS program creates conditions which are likely to place anthropologists in positions in which their work will be in violation of the AAA Code of Ethics and . . . its use of anthropologists poses a danger to both other anthropologists and persons other anthropologists study. Thus the Executive Board expresses its disapproval of the HTS program."[8] The involvement of psychologists, psychiatrists, and other medical personnel with military and intelligence programs that employ harsh methods of interrogation, including waterboarding and other brutal measures deemed by many critics to constitute torture, has likewise prompted probing criticism of the nexus connecting professors to national security agencies in the post-9/11 era.[9]

*****

From the mid-1940s to the early 1960s, patrons and their close social science collaborators promoted a widespread strategy that emphasized the place of the social sciences within a unified scientific enterprise, that assumed their progress depended on the cultivation of social science inquiry with particular characteristics that made it distinct from humanistic studies, value-laden discourses, and ideological positions, and that conceived of basic studies and more practically oriented research in ways that would lead to great benefits. Though that era was not so long ago, much has changed since then. The political, institutional, and intellectual climates that had encouraged social scientists and their patrons to advocate those ideas so widely no longer exist. Nowadays discussions about the disunity of the sciences proliferate. Arguments that the social sciences and even individual social science disciplines do not offer anything resembling a unified investigative approach compete favorably with the alternative. To hear that particular research projects and even major lines of social and psychological inquiry reflect and reinforce a certain political or ideological orientation hardly surprises anyone. And to wonder if such research might reflect the interests of its patrons seems nothing but commonsensical. Widespread acceptance of the claim that social science progress depends on adopting a unity-of-science viewpoint and corresponding assumptions about the apolitical and value-neutral character of legitimate scientific inquiry seems to be long gone and not likely to return any time soon.

On the other hand, large numbers of academic scholars remain committed to the notion that their fields of social research should proceed in a manner that is as objective as possible, in order to ensure that the resulting knowledge is unbiased and thus recognized as trustworthy by reasonable people regardless of their particular ideological or political viewpoint. Even researchers who accept funding from partisan foundations often claim that their research has objective scientific validity, meaning that their findings have not been compromised or contaminated by the political or ideological orientation of their sponsor, or even by their own personal commitments. And as a society we are still influenced in a myriad of ways by the findings, recommendations, perspectives, and techniques of analysis from the social, psychological, and behavioral sciences. Do we have good grounds for believing that these researchers' entanglements with patrons do not hopelessly undermine the validity of their work?

Understanding how we have gotten to this polyvalent and even deeply confusing historical juncture requires careful consideration of the changing funding landscape. We need to follow the money in order to understand what gets studied, by whom, how, under what conditions, and for what purposes. To be sure, it would be going too far to claim that those who pay the piper always call the tune, for sponsors of research do not always get what they ask for. Furthermore, such a claim about the unbounded power of funders would fail to acknowledge the roles of individual social scientists, scholarly organizations, universities, research institutes, and other interested parties from the political, business, scientific, and international communities that have shaped patron interests and conditioned their engagements with the social sciences. But to insist on these points does not suggest the diminished importance of patrons. Rather, it indicates that we also need to pay careful attention to the complex and changing force fields within which patrons operate. To understand the intellectual commitments of the social sciences, their professional outlook, and their influence on society and public policy, we need to pay careful attention to their changing relationships with patrons and associated developments in the politics–patronage–social science nexus.

NOTES

INTRODUCTION

1. Membership numbers from the Behavioral and Social Sciences Survey Committee of the National Academy of Sciences and the Social Science Research Council, *The Behavioral and Social Sciences: Outlook and Needs* (Englewood Cliffs, NJ: Prentice-Hall, 1969), 23.
2. Roger E. Backhouse and Philippe Fontaine, eds., *The History of the Social Sciences since 1945* (New York: Cambridge Univ. Press, 2010); Mark Solovey and Hamilton Cravens, eds., *Cold War Social Science: Knowledge Production, Liberal Democracy, and Human Nature* (New York: Palgrave Macmillan, 2012).
3. The following books and historiographic essays provide useful points of entry into this vibrant field of study: Christopher Simpson, *Science of Coercion: Communication Research and Psychological Warfare, 1945–1960* (New York: Oxford Univ. Press, 1994); Ellen Herman, *The Romance of American Psychology: Political Culture in the Age of Experts* (Berkeley: Univ. of California Press, 1995); Christopher Simpson, ed., *Universities and Empire: Money and Politics in the Social Sciences during the Cold War* (New York: New Press, 1998); James H. Capshew, *Psychologists on the March: Science, Practice, and Professional Identity in America, 1929–1969* (New York: Cambridge Univ. Press, 1999); Ron Robin, *The Making of the Cold War Enemy: Culture and Politics in the Military-Intellectual Complex* (Princeton, NJ: Princeton Univ. Press, 2001); Philip Mirowski, *Machine Dreams: Economics Becomes a Cyborg Science* (New York: Cambridge Univ. Press, 2001); Hunter Crowther-Heyck, *Herbert A. Simon: The Bounds of Reason in Modern America* (Baltimore: Johns Hopkins Univ. Press, 2005); Hunter Crowther-Heyck, "Patrons of the Revolution: Ideals and Institutions in Postwar Behavioral Science," *Isis* 97 (2006), 420–446; Joel Isaac, "The Human Sciences in Cold War America," *The Historical Journal* 50 (2007), 725–746; David C. Engerman, *Know Your Enemy: The Rise and Fall of America's Soviet Experts* (New York: Oxford Univ. Press, 2009); Matthew Farish, *The Contours of America's Cold War* (Minneapolis: Univ. of Minnesota Press, 2010); David C. Engerman, "Social Science in the Cold War," *Isis* 101 (2010), 393–400. Also see the works cited in note 2, by Backhouse and Fontaine, and by Solovey and Cravens.
4. Important exceptions include Simpson, *Science of Coercion*; Alice O'Connor, *Poverty Knowledge: Social Science, Social Policy, and the Poor in Twentieth-Century U.S. History* (Princeton, NJ: Princeton Univ. Press, 2001); Robin, *Making of the Cold War Enemy*; and Engerman, *Know Your Enemy*.
5. For example, see these two well-known edited volumes about universities, the social sciences (though not exclusively), and the Cold War: Simpson, *Universities and Empire*; Noam Chomsky et al., eds., *The Cold War and the University: Toward an Intellectual History of the Postwar Years* (New York: New Press, 1997).

6. Exceptions include Hunter Crowther-Heyck, "Herbert Simon and the GSIA: Building an Interdisciplinary Community," *Journal of the History of the Behavioral Sciences* 42 (2006), 311–334; Mark Solovey and Jefferson D. Pooley, "The Price of Success: Sociologist Harry Alpert, the NSF's First Social Science Policy Architect," *Annals of Science* 68 (2011), 229–260.
7. For the history of Western social science, see H. Scott Gordon, *The History and Philosophy of Social Science* (New York: Routledge, 1991); Roger Smith, *The Norton History of the Human Sciences* (New York: W. W. Norton, 1997); Theodore M. Porter and Dorothy Ross, eds., *The Cambridge History of Science*, vol. 7: *The Modern Social Sciences* (New York: Cambridge Univ. Press, 2003). For the history of American social science, see Peter T. Manicas, *A History and Philosophy of the Social Sciences* (New York: Basil Blackwell, 1987); Dorothy Ross, *Origins of American Social Science* (New York: Cambridge Univ. Press, 1991). For the discipline of history, see Peter Novick, *That Noble Dream: The 'Objectivity Question' and the American Historical Profession* (New York: Cambridge Univ. Press, 1988).
8. Ellen C. Lagemann, *The Politics of Knowledge: The Carnegie Corporation, Philanthropy, and Public Policy* (Middletown, CT: Wesleyan Univ. Press, 1989); Donald Fisher, *Fundamental Development of the Social Sciences: Rockefeller Philanthropy and the United States Social Science Research Council* (Ann Arbor: Univ. of Michigan Press, 1993); Richard S. Kirkendall, *Social Scientists and Farm Politics in the Age of Roosevelt* (Columbia: Univ. of Missouri Press, 1966).
9. See Simpson, *Science of Coercion*; Jennifer Platt, *A History of Sociological Research Methods in America, 1920–1960* (New York: Cambridge Univ. Press, 1996); Michael A. Bernstein, *A Perilous Progress: Economists and Public Purpose in Twentieth-Century America* (Princeton, NJ: Princeton Univ. Press, 2001); Robin, *Making of the Cold War Enemy*.
10. Though there were other important funding sources during this period, they tended to concentrate more narrowly on a particular discipline or area of research than the ones I have included in this study. For example, see Wade E. Pickren and Stanley F. Schneider, eds., *Psychology and the National Institute of Mental Health: A Historical Analysis of Science, Practice, and Policy* (Washington, DC: American Psychological Association, 2004).
11. Fisher, *Fundamental Development of the Social Sciences*; Thomas F. Gieryn, *Cultural Boundaries of Science: Credibility on the Line* (Chicago: Univ. of Chicago Press, 1999); Charles A. Taylor, *Defining Science: A Rhetoric of Demarcation* (Madison: Univ. of Wisconsin Press, 1996).
12. Manicas, *A History and Philosophy of the Social Sciences*; Ross, *Origins of American Social Science*; Smith, *The Norton History of the Human Sciences*.
13. Teller remark recounted in Theda Skocpol, "Governmental Structures, Social Science, and the Development of Economic and Social Policies," 40–50, in Martin Bulmer, ed., *Social Science Research and Government: Comparative Essays on Britain and the United States* (New York: Cambridge Univ. Press, 1987), 40.
14. On this time period, see Mary O. Furner, *Advocacy and Objectivity: A Crisis in the Professionalization of American Social Science, 1865–1905* (Lexington: Univ. Press of Kentucky, 1975); Thomas L. Haskell, *The Emergence of Professional Social Science: The American Social Science Association and the Nineteenth-Century Crisis of Authority* (Urbana: Univ. of Illinois Press, 1977); Ross, *Origins of American Social Science*; John M. Jordan, *Machine-Age Ideology: Social Engineering and American Liberalism, 1911–1939* (Chapel Hill: Univ. of North Carolina Press, 1994); Julie A. Reuben, *The Making of the Modern University: Intellectual Transformation and the Marginalization of Morality* (Chicago: Univ. of Chicago Press, 1996).

15. On the SSRC and the Rockefeller philanthropies, see Fisher, *Fundamental Development of the Social Sciences*; Martin Bulmer and Joan Bulmer, "Philanthropy and Social Science in the 1920s: The Case of Beardsley Ruml and the Laura Spelman Rockefeller Memorial, 1922–29," *Minerva* 19 (1981): 347–407. On the Carnegie Corporation, see Lagemann, *The Politics of Knowledge*. On the Brookings Institution, see Donald T. Critchlow, *The Brookings Institution, 1916–1952: Expertise and the Public Interest in a Democratic Society* (DeKalb: Northern Illinois Univ. Press, 1985). Many disciplinary histories deal with these issues as well, for example, Robert C. Bannister, *Sociology and Scientism: The American Quest for Objectivity, 1880–1940* (Chapel Hill: Univ. of North Carolina Press, 1987).
16. Ross, *Origins of American Social Science*, 1929, at 471. Ross's argument is interesting also because she links the ideology of American exceptionalism to the advance of scientism.
17. Isaac, "The Human Sciences in Cold War America," 740. On a similar note, David Engerman, in "Social Science in the Cold War," has recently urged historians to consider how developments in the social sciences during the Cold War years were rooted in earlier events and debates, including the experiences of social scientists during World War Two.
18. See especially Mark C. Smith, *Social Science in the Crucible: The American Debate over Objectivity and Purpose, 1918–1941* (Durham, NC: Duke Univ. Press, 1994). The cases of psychology, economics, political science, and history have all received extensive analysis. On psychology: Katherine Pandora, *Rebels within the Ranks: Psychologists' Critique of Scientific Authority and Democratic Realities in New Deal America* (New York: Cambridge Univ. Press, 1997); Ian A. M. Nicholson, *Inventing Personality: Gordon Allport and the Science of Selfhood* (Washington, DC: American Psychological Association, 2003). On economics: Yuval P. Yonay, *The Struggle over the Soul of Economics: Institutional and Neoclassical Economics in America between the Wars* (Princeton, NJ: Princeton Univ. Press, 1998); Mary S. Morgan and Malcolm Rutherford, eds., *From Interwar Pluralism to Postwar Neoclassicism*, annual supplement to *History of Political Economy*, vol. 30 (Durham, NC: Duke Univ. Press, 1999). On political science: John G. Gunnell, *The Descent of Political Theory: The Genealogy of an American Vocation* (Chicago: Univ. of Chicago Press, 1993). On history: Novick, *That Noble Dream*. Landmark primary source materials include Robert S. Lynd, *Knowledge for What? The Place of Social Science in American Culture* (Princeton, NJ: Princeton Univ. Press, 1939), and Robert M. Hutchins, *The Higher Learning in America*, with a new introduction by Harry S. Ashmore (New Brunswick, NJ: Transaction Publishers, 1995/1936).
19. Patrick D. Reagan, *Designing a New America: The Origins of New Deal Planning, 1890–1943* (Amherst: Univ. of Massachusetts Press, 2000); Kirkendall, *Social Scientists and Farm Politics in the Age of Roosevelt*; William J. Barber, *Designs within Disorder: Franklin D. Roosevelt, the Economists, and the Shaping of American Economic Policy, 1933–1945* (New York: Cambridge Univ. Press, 1996).
20. Prewitt, who served as the U.S. Census Bureau director, states "The US social sciences started as and have continued to be American-centric. They have overwhelmingly focused their energies on issues of American polity, economy, and society." See Prewitt, "The Two Projects of the American Social Sciences," *Social Research* 72 (2005), 1–20, quotation at 9.
21. Though the following studies do not invoke this term explicitly, they show how the general idea can be fruitfully developed through carefully contextualized historical accounts: Michael E. Latham, *Modernization as Ideology: American Social Science and "Nation Building" in the Kennedy Era* (Chapel Hill: Univ. of North Carolina Press, 2000);

Nils Gilman, *Mandarins of the Future: Modernization Theory in Cold War America* (Baltimore: Johns Hopkins Univ. Press, 2003); Sarah Igo, *The Averaged American: Surveys, Citizens, and the Making of a Mass Public* (Cambridge, MA: Harvard Univ. Press, 2007). On the coproduction of science and society, see Sheila Jasanoff, ed., *States of Knowledge: The Co-Production of Science and Social Order* (New York: Routledge, 2004).

22. Mark Solovey, "Project Camelot and the 1960s Epistemological Revolution: Rethinking the Politics–Patronage–Social Science Nexus," *Social Studies of Science* 31 (2001), 171–206.
23. Historian of science Allan A. Needell has written that "the potential impact of such support [from military and intelligence agencies]—direct or indirect—on the quality and independence of research and on the teaching of these subjects remained largely unevaluated," at least until the 1960s: "'Truth Is Our Weapon': Project TROY, Political Warfare, and Government-Academic Relations in the National Security State," *Diplomatic History* 17 (1993), 399–420, quotation at 418.
24. Elsewhere, I have suggested that in the large body of scholarship on science in Cold War America, the blurring of boundaries between science and politics has been a prominent theme of analysis as well as a major source of anxiety: Mark Solovey, "Introduction: Science and the State during the Cold War: Blurred Boundaries and a Contested Legacy," *Social Studies of Science* 31 (2001), 165–170.
25. Friedrich A. Hayek, *Individualism and Economic Order* (Chicago: Univ. of Chicago Press, 1948); Hayek, *The Counter-Revolution of Science: Studies on the Abuse of Reason* (Glencoe, IL: Free Press, 1952). For a recent historical discussion, see Naomi Beck, "In Search of the Proper Scientific Approach: Hayek's Views on Biology, Methodology, and the Nature of Economics," *Science in Context* 22 (2009), 567–585. To the present day, scientism remains a term of abuse used to denounce the inappropriate extension of attitudes, concepts, methods, and goals from the natural sciences to some other field of study. See, for example, Tom Sorell, *Scientism: Philosophy and the Infatuation with Science* (New York: Routledge, 1991); Susan Haack, *Defending Science—within Reason: Between Science and Cynicism* (New York: Prometheus Books, 2007).
26. Richard Olson, *Science and Scientism in Nineteenth-Century Europe* (Urbana: Univ. of Illinois Press, 2003).
27. On the various terms used to define this field, their meanings, and what the controversy over terms and meanings implies for how we should write our histories, see Theodore M. Porter and Dorothy Ross, "Introduction: Writing the History of Social Science," 1–10, in Porter and Ross, *The Cambridge History of Science*, vol. 7; Roger Smith, "History and the History of the Human Sciences: What Voice?" *History of the Human Sciences* 10 (1997), 22–39.
28. Stuart Chase, *The Proper Study of Mankind*, rev. ed. (New York: Harper & Bros., 1956/1948), 129.

CHAPTER 1. SOCIAL SCIENCE ON THE ENDLESS (AND END-LESS?) FRONTIER

The epigraphs to chapter 1 are drawn from George A. Lundberg, "The Senate Ponders Social Science," *Scientific Monthly* 64 (1947), 397–411, quotation at 409–410; Louis Wirth, "Responsibility of Social Science," *Annals of the American Academy of Political and Social Science* 249 (1947), 143–151, quotation at 148.

1. Paul K. Hoch, "The Crystallization of a Strategic Alliance: The American Physics Elite and the Military in the 1940's," in Everett Mendelsohn, Merritt Roe Smith, and Peter

Weingart, eds., *Science, Technology and the Military* (Boston: Kluwer, 1988), 87–116, dollar figures on 96.

2. J. Merton England, *A Patron for Pure Science: The National Science Foundation's Formative Years, 1945–57* (Washington, DC: NSF, 1982); Daniel L. Kleinman, *Politics on the Endless Frontier: Postwar Research Policy in the United States* (Durham, NC: Duke Univ. Press, 1995); Jessica Wang, "Liberals, the Progressive Left, and the Political Economy of Postwar American Science: The National Science Foundation Debate Revisited," *Historical Studies in the Physical and Biological Sciences* 26 (1995), 139–166; Jessica Wang, *American Science in an Age of Anxiety: Scientists, Anticommunism, and the Cold War* (Chapel Hill: Univ. of North Carolina Press, 1999).
3. Some social scientists and mainly sociologists have discussed the importance of this episode: Gene M. Lyons, *The Uneasy Partnership: Social Science and the Federal Government in the Twentieth Century* (New York: Russell Sage Foundation, 1969), 126–136; Samuel Z. Klausner and Victor M. Lidz, eds., *The Nationalization of the Social Sciences* (Philadelphia: Univ. of Pennsylvania Press, 1986); Otto N. Larsen, *Milestones and Millstones: Social Science at the National Science Foundation, 1945–1991* (New Brunswick, NJ: Transaction, 1992), 1–18; Thomas F. Gieryn, "The U.S. Congress Demarcates Natural Science and Social Science (Twice)," 65–114, in Gieryn, *Cultural Boundaries of Science: Credibility on the Line* (Chicago: Univ. of Chicago Press, 1999); Uta Gerhardt, *Talcott Parsons: An Intellectual Biography* (New York: Cambridge Univ. Press, 2002), 149–167. However, historians of the social sciences have paid only brief attention: Michael A. Bernstein, *A Perilous Progress: Economists and Public Purpose in Twentieth-Century America* (Princeton, NJ: Princeton Univ. Press, 2001), 100–101; James H. Capshew, *Psychologists on the March: Science, Practice, and Professional Identity in America, 1929–1969* (New York: Cambridge Univ. Press, 1999), 176–179; Ellen Herman, *The Romance of American Psychology: Political Culture in the Age of Experts* (Berkeley: Univ. of California Press, 1995), 46. For a more substantial account, see my essay: Solovey, "Riding Natural Scientists' Coattails onto the Endless Frontier: The SSRC and the Quest for Scientific Legitimacy," *Journal of the History of the Behavioral Sciences* 40 (2004), 393–422.
4. Robert F. Maddox, *The Senatorial Career of Harley Martin Kilgore* (New York: Garland, 1981), 162–173, 329–330.
5. Vannevar Bush, *Science—The Endless Frontier* (Washington, DC: NSF, 1995, 50th anniversary reissue of 1945 edition), 9 (hereafter, Bush, *SEF*). Also see G. Pascal Zachary, *Endless Frontier: Vannevar Bush, Engineer of the American Century* (New York: Free Press, 1997), 218–239, 249–260; Nathan Reingold, "Vannevar Bush's New Deal for Research; or, The Triumph of the Old Order," in Reingold, *Science, American Style* (New Brunswick, NJ: Rutgers Univ. Press, 1991), 284–333.
6. Bush, *SEF*, 9, 12.
7. U.S. Congress, Senate, Committee on Military Affairs, Subcommittee on War Mobilization, *Legislative Proposals for the Promotion of Science: The Texts of Five Bills and Excerpts from Reports*, 79th Cong., 1st sess. (Washington, DC: U.S. GPO, 1945).
8. Bush, *SEF*, 23. Bush to Truman, July 5, 1945, in Bush, *SEF*, 1.
9. Ogburn presented this account to NSF's first social science policy architect, Harry Alpert, as noted in Harry Alpert to Files, July 30, 1953, document in author's possession.
10. Vannevar Bush to D. C. Josephs, Sept. 19, 1946, folder Early Historical Documents 1943–1953, #2, NSF Historian's File (hereafter, NSF HF). The archival collection I designate NSF HF was located at the NSF headquarters in Arlington, VA, when I did my research. Subsequently, the materials from this collection were integrated into Record Group 307 (RG 307), Records of the National Science Foundation, National Archives. In some

cases but not all, the folders from the NSF HF along with their contents were maintained intact during the process of integration into RG 307.

11. Dorothy McLean, "Anthropology and Psychology: The Borderland Division of the National Research Council," Mar. 12, 1954, p. 1, folder A & P: "Borderland Division of NAS" 1954, Anthropology and Psychology Series, NAS Archives, courtesy of National Academy of Sciences Archives. Also see Rexmond C. Cochrane, *National Academy of Sciences: The First Hundred Years: 1863–1963* (Washington DC: NAS, 1978).
12. Charles E. Merriam, "The National Resources Planning Board: A Chapter in American Planning Experience," *American Political Science Review* 38 (1944), 1075–1088; Cochrane, *National Academy of Sciences*, 347–381.
13. Talcott Parsons, "The Science Legislation and the Role of the Social Sciences," *American Sociological Review* 11 (1946), 653–666, esp. 658–659.
14. Charles W. Bray, *Psychology and Military Proficiency: A History of the Applied Psychology Panel of the National Defense Research Committee* (Princeton, NJ: Princeton Univ. Press, 1948).
15. U.S. NRPB, *Post-War Agenda* (Washington, DC: U.S. GPO, 1942); Patrick D. Reagan, *Designing a New America: The Origins of New Deal Planning, 1890–1943* (Amherst: Univ. of Massachusetts Press, 1999).
16. Robert K. Merton, "The Role of the Intellectual in the Public Bureaucracy," *Social Forces* 23 (1945), 405–415, quotation at 408.
17. Daniel J. Kevles, *The Physicists: The History of a Scientific Community in Modern America* (New York: Knopf, 1977), 345. Charlatans in Vannevar Bush to D. C. Josephs, Oct. 10, 1946, folder SSRC—Overall Survey of the Nature and Needs of the Social Sciences (Donald Marquis), box 2 (SSRC), Carnegie Corporation of New York Records, Columbia University Butler Library, New York, NY.
18. Reminiscences of Vannevar Bush, 1967, Carnegie Corporation Project, pp. 47–48, in the Columbia Center for Oral History, Columbia University Libraries, New York, NY.
19. Bush's testimony in U.S. Congress, Senate, A Subcommittee of the Committee on Military Affairs, *Hearings on Science Legislation (S. 1297 and Related Bills)*, 79th Cong., 1st sess., pts. 1–5 (Washington, DC: U.S. GPO, 1945) (hereafter, *1945 Senate hearings*), 200; also U.S. Congress, House, A Subcommittee of the Committee on Interstate and Foreign Commerce, *Hearings on National Science Foundation Act*, 79th Cong., 2d sess., (Washington, DC: U.S. GPO, 1946) (hereafter, *1946 House hearings*), 53.
20. Excerpt from Irvin Stewart to A. P. Brogan, Sept. 24, 1945, folder Early Historical Documents 1943–1953, #1, NSF HF. Harry S. Truman, "Special Message to the Congress Presenting a 21-Point Program for the Reconversion Period, Sept. 6, 1945," in *Public Papers of the Presidents, Truman, 1945* (Washington, DC: U.S. GPO, 1962), 292–294. Smith quoted in Reingold, "Vannevar Bush's New Deal for Research," 311.
21. President Roosevelt, "Let Us Move Forward with Strong and Active Faith," undelivered Jefferson Day address, Apr. 13, 1945, in Samuel I. Rosenman, ed., *The Public Papers and Addresses of Franklin D. Roosevelt, 1944–1945* (New York: Harper, 1950), 613–616, quotation at 615.
22. Larsen, *Milestones and Millstones*, 6.
23. For example, see Gieryn, "The U.S. Congress Demarcates Natural Science and Social Science (Twice)," 67–68.
24. On the SSRC, see Donald Fisher, *Fundamental Development of the Social Sciences: Rockefeller Philanthropy and the United States Social Science Research Council* (Ann Arbor: Univ. of Michigan Press, 1993).
25. Ibid., 191.

26. *SSRC Annual Report 1944–45*, 7.
27. Louis Wirth, *Report on the History, Activities and Policies of the Social Science Research Council*, unpublished, Aug. 1937, p. 151, folder 2, box 32, Wirth Papers, Special Collections Research Center, University of Chicago Library (hereafter, Wirth Papers).
28. C. Wright Mills, "The Social Role of the Intellectual," in Mills, *Power, Politics and People: The Collected Essays of C. Wright Mills*, ed. Irving L. Horowitz (New York: Oxford Univ. Press, 1963), 292–304, quotations at 297, 302.
29. The Committee on Science and the Federal Government, "The Federal Government and Research in the Social Sciences," Apr. 14, 1944, p. 2, folder 1894, box 168, subseries 37, series 1, RG 2, Social Science Research Council Archives, Rockefeller Archive Center, Sleepy Hollow, NY (hereafter, SSRC Archives).
30. Ibid., 2.
31. The position of Mitchell's committee presented in Minutes, Committee on Problems and Policy, Apr. 14, 1945, p. 1, folder 1785, box 315, subseries 1, series 2, RG 1, SSRC Archives, and appendix 1, "The Federal Government and Research," pp. 2, 3, attached to these same minutes.
32. Minutes, Committee on Problems and Policy, Apr. 14, 1945, pp. 3, 5.
33. Minutes, Committee on Problems and Policy, July 28–29, 1945, pp. 1, 2, folder 1785, box 315, subseries 1, series 2, RG 1, SSRC Archives.
34. Minutes, Board of Directors Meeting, Sept. 10–13, 1945, gravest at 7, Hauser at 9–10, folder 2098, box 357, series 9, RG 1, SSRC Archives.
35. The position of Mitchell's Committee is reported in ibid., 8.
36. On foundation funding, see chapter 3. As of 1945, the best-known case of industrial support involved the famous Hawthorne experiments, which promised to help increase worker productivity and diminish labor unrest. See Richard Gillespie, *Manufacturing Knowledge: A History of the Hawthorne Experiments* (New York: Cambridge Univ. Press, 1991). On military funding, see chapter 2.
37. Parsons, "The Science Legislation and the Role of the Social Sciences," 660.
38. Mitchell's Committee in appendix 16, draft "Report of the Committee on the Federal Government and Research," Sept. 10, 1945, p. 8, folder 863, box 154, subseries 19, series 1, RG 1, SSRC Archives. Leland in Minutes, Board of Directors Meeting, Sept. 10–13, 1945, p. 17.
39. Bacon quoted on p. 187 in Robert K. Merton and Paul K. Hatt, "Election Polling Forecasts and Public Images of Social Science: A Case Study in the Shaping of Opinion among a Strategic Public," *Public Opinion Quarterly* 13 (1949), 185–222.
40. On the distribution of witness opinions, see U.S. Congress, House, *Technical Information for Congress*, Report to the Subcommittee on Science, Research, and Development of the Committee on Science and Astronautics, 92d Cong., 1st sess., Serial A (Washington, DC: U.S. GPO, Apr. 25, 1969, rev. May 1, 1971), 113.
41. Bush in *1945 Senate hearings*, 200.
42. Bush, *SEF*, 18, 19.
43. In *1945 Senate hearings*: Chemical Society's position reported by Roger Adams at 806–807, Bakhmeteff at 715, Dewey at 818.
44. Ibid., Compton at 631, Rabi at 998–999.
45. Isaiah Bowman, *Geography in Relation to the Social Sciences* (New York: Charles Scribner's Sons, 1934), 227; Bowman in *1945 Senate hearings*, 23; Bowman on "labor" quoted in England, *Patron for Pure Science*, 49; Neil Smith, *American Empire: Roosevelt's Geographer and the Prelude to Globalization* (Los Angeles: Univ. of California Press, 2003), 435.

On the wider history of scientific objectivity, see Lorraine J. Daston and Peter Galison, *Objectivity* (Zone Books, 2007).

46. In *1945 Senate hearings*: Rabi at 998–999, Fishbein at 496, Bakhmeteff at 715, Simms at 1170.
47. Rabi in *1945 Senate hearings*, 998–999.
48. Warren G. Magnuson to the SSRC, Sept. 19, 1945, p. 2, folder 1894, box 168, subseries 37, series 1, RG 2, SSRC Archives. On the APA resolution, see SSRC Minutes, Board of Directors Meeting, Sept. 10–13, p. 16. On the ASS resolution, see Louis Wirth to Talcott Parsons (and others) Mar. 6, 1947, folder 6, box 17, Wirth Papers. On support from other professional societies, see SSRC Minutes, Board of Directors Meeting, July 20–21, 1946, p. 3, folder 1787, box 316, subseries 1, series 2, RG 1, SSRC Archives.
49. Gieryn, "The U.S. Congress Demarcates Natural Science and Social Science (Twice)," 73.
50. Wesley C. Mitchell, "Empirical Research and the Development of Economic Science," 3–20, in NBER, *Economic Research and the Development of Economic Science and Public Policy* (New York: NBER, 1946), 10. On Mitchell, see Mark C. Smith, *Social Science in the Crucible: The American Debate over Objectivity and Purpose, 1918–1941* (Durham, NC: Duke Univ. Press, 1994), 49–63.
51. William F. Ogburn, "Science and Society," in Robert C. Stauffer, ed., *Science and Civilization* (Madison: Univ. of Wisconsin Press, 1949), 197–212, quotation at 205; William F. Ogburn, "The Folkways of a Scientific Sociology," *The Scientific Monthly* 30 (1930), 300–306, quotation at 301. On Ogburn, see Robert C. Bannister, *Sociology and Scientism: The American Quest for Objectivity, 1880–1940* (Chapel Hill: Univ. of North Carolina Press, 1987), 161–187.
52. Edwin G. Nourse, "Economic Analysis and Political Synthesis," 1950, reprinted as appendix F in Nourse, *Economics in the Public Service: Administrative Aspects of the Employment Act* (New York: Harcourt, Brace, 1953), 496–503, quotation at 500. On Nourse, see Joseph G. Knapp, *Edwin G. Nourse—Economist for the People* (Danville, IL: Interstate Printers & Publishers, 1979).
53. Robert M. Yerkes, "The Scope of Science," *Science* 105 (May 2, 1947), 461–463, quotation at 463. On Yerkes, see Capshew, *Psychologists on the March*, 42–51.
54. John M. Gaus, *Reflections on Public Administration* (Tuscaloosa: Univ. of Alabama Press, 1947).
55. Nevertheless, historians followed the debate: Louis Knott Koontz, "The Social Sciences in the National Science Foundation," *Pacific Historical Review* 15 (1946), 1–30.
56. Ogburn in *1945 Senate hearings*, 769.
57. Ibid., Ogburn at 773, Mitchell at 739.
58. Ibid., Gaus at 747, Mitchell at 741, Nourse at 757.
59. Ibid., Nourse at 757, 758, Yerkes at 755, 743.
60. Regarding the agency's director, see Kilgore's exchange with Mitchell and Ogburn in ibid., 785–786.
61. See Wang, "Liberals, the Progressive Left, and the Political Economy of Postwar American Science," and her book *American Science in an Age of Anxiety*.
62. In *1945 Senate hearings*: Kilgore at 22, 632; Wallace at 137, 140; Magnuson at 51–52.
63. Bowman Committee letter in ibid., 1126–1129.
64. "Original Members of the Committee for a National Science Foundation," attached to Harlow Shapley and Harold C. Urey to Dear Member, July 18, 1946, folder 8, box 30, Wirth Papers.

65. U.S. Congress, Senate, Subcommittee on War Mobilization of the Committee on Military Affairs, *National Science Foundation, Report on Science Legislation,* 79th Cong., 2d sess. (Washington, DC: U.S. GPO, Feb. 27, 1946), 6. The Committee on Military Affairs quoted in *Technical Information for Congress,* 114. Kilgore in *Congressional Record—Senate,* July 3, 1946, 8231–8232.
66. Carl C. Taylor, "The Social Responsibilities of the Social Sciences—The National Level," *American Sociological Review* 11 (1946), 384–392, quotation at 386; Memorandum, undated and anonymous, "Crippling the Work of Social Scientists in the U.S. Bureau of Agricultural Economics, by Means of Riders Attached to Appropriation Bills," folder 5, box 18, Wirth Papers. On the BEA, see Richard S. Kirkendall, *Social Scientists and Farm Politics in the Age of Roosevelt* (Columbia: Univ. of Missouri Press, 1966).
67. U.S. NRPB, *After the War—Full Employment* (Washington, DC: U.S. GPO, 1942); William J. Barber, *Designs within Disorder: Franklin D. Roosevelt, the Economists, and the Shaping of American Economic Policy, 1933–1945* (New York: Cambridge Univ. Press, 1996); Stephen K. Bailey, *Congress Makes a Law: The Story Behind the Employment Act of 1946* (New York: Columbia Univ. Press, 1950). The 1946 enabling legislation also created the Council of Economic Advisers, which immediately became involved in the enduring conflict between advocacy and neutrality in social science, between the ideals of social engagement and scientific detachment.
68. Gunnar Myrdal, with the assistance of Richard Sterner and Arnold Rose, *An American Dilemma: The Negro Problem and Modern Democracy* (New York: Harper, 1944); Walter A. Jackson, *Gunnar Myrdal and America's Conscience: Social Engineering and Racial Liberalism, 1938–1987* (Chapel Hill: Univ. of North Carolina Press, 1990).
69. Leo Bogart, "Introduction," 1–41, in Bogart, ed., *Social Research and the Desegregation of the U.S. Army: Two Original 1951 Field Reports* (Chicago: Markham, 1969). "The Federal Government and Research in the Social Sciences," Apr. 14, 1944, quotation at 4.
70. Taft in *Congressional Record—Senate,* July 2, 1946, p. 8145. On Taft, see James T. Patterson, *Mr. Republican: A Biography of Robert A. Taft* (Boston: Houghton Mifflin, 1972). Smith in *Congressional Record—Senate,* July 3, 1946, p. 8231.
71. Hart in *Congressional Record—Senate,* July 3, 1946, pp. 8230, 8232.
72. In *1946 House hearings*: Mills at 3, Bowman at 14, Brown at 13.
73. *Congressional Record—Senate,* May 14, 1947: Smith at 5258, Fulbright on 5511–5512. On the 1947 compromise bill, see Kleinman, *Politics on the Endless Frontier,* 132–133.
74. Harold Orlans, "Academic Social Scientists and the Presidency: From Wilson to Nixon," *Minerva* 24 (1986), 172–204, Truman's quip at 186.
75. President's Scientific Research Board (John R. Steelman, chair), *Science and Public Policy,* vol. 1 (Washington, DC: U.S. GPO, 1947), viii.
76. Fulbright in *Congressional Record—Senate,* May 4, 1948, p. 5251.
77. Minutes, Board of Directors Meeting, July 20–21, 1946, p. 4, folder 1787, box 316, subseries 1, series 2, RG 1, SSRC Archives.
78. Wirth's best-known writings include his one book *The Ghetto* (Chicago: Univ. of Chicago Press, 1928), and "Urbanism as a Way of Life," *American Journal of Sociology* 44 (1938), 1–24. Roger A. Salerno, *Louis Wirth: A Bio-Bibliography* (New York: Greenwood Press, 1987).
79. Louis Wirth, "The Unfinished Business of American Democracy," *Annals of the American Academy of Political and Social Science* 244 (1946), 1–9, quotations at 1, 2.
80. Wirth, preface to Karl Mannheim, *Ideology and Utopia: An Introduction to the Sociology of Knowledge* (London: Routledge, 1936), xxvii–xxviii. David Kettler and Volker Meja, *Karl Mannheim and the Crisis of Liberalism: The Secret of These New Times* (New Brunswick,

NJ: Transaction, 1995), 234–235. Myrdal, *An American Dilemma,* esp. appendix 2, "A Methodological Note on Facts and Valuations in Social Science," 1035–1064, quotation at 1045.

81. Louis Wirth to E. W. Burgess, May 29, 1946, folder 1895, box 168, subseries 37, series 1, RG 2, SSRC Archives; Wirth, "Responsibility of Social Science," 148.
82. John Dewey, "Liberating the Social Scientist," *Commentary* 4 (1947), 378–385; Charles A. Beard, "Neglected Aspects of Political Science," *American Political Science Review* 42 (1948), 211–222; Robert S: Lynd, "The Science of Inhuman Relations," *New Republic* 121 (Aug. 29, 1949), 22–25; Louis Wirth, "Karl Mannheim, 1893–1947," *American Sociological Review* 12 (1947), 356–357, and Louis Wirth, undated, "On Making Values Explicit," folder 9, box 57, Wirth Papers; Gunnar Myrdal, *The Political Element in the Development of Economic Theory*, trans. from the German by Paul Streeten (Cambridge, MA: Harvard Univ. Press, 1954).
83. Sub-Committee on Social Science and Values, Minutes, Meeting of Feb. 9, 1949, folder 1369, box 226, subseries 19, series 1, RG 1, SSRC Archives, Wirth at 1, 2.
84. Ibid., 7–8.
85. Louis Wirth, "Comments on Social Sciences and Values," Sept. 1949, folder 1370, box 226, subseries 19, series 1, RG 1, SSRC Archives. Appendix, "Value Problems and Social Science Research, Progress Reports on the University Seminars," attached to Minutes, Problem and Policy Committee, Sept. 10–11, 1949, methods at 5, folder 1369, box 226, subseries 19, series 1, RG 1, SSRC Archives.
86. SSRC Minutes, Committee on Problems and Policy, July 20–21, 1946, p. 4.
87. Ibid.; Waldemar Kaempffert, *Should the Government Support Science?* (Public Affairs Committee, Inc., 1946).
88. SSRC Committee on the Federal Government and Research, Minutes, Nov. 15, 1946, p. 3, folder 864, box 154, subseries 19, series 1, RG 1, SSRC Archives.
89. Parsons's respect for Wirth is evident in Parsons's detailed response to Wirth's review of his 1937 book: Talcott Parsons to Louis Wirth, Oct. 6, 1939, folder 9, box 9; Louis Wirth to Talcott Parsons, Nov. 13, 1937, folder 9, box 9; Louis Wirth to Talcott Parsons (and others), Mar. 6, 1947, folder 6, box 17. All three letters in Wirth Papers.
90. Parsons's two major works were *The Structure of Social Action: A Study in Social Theory with Special Reference to a Group of Recent European Writers* (New York: McGraw Hill, 1937), and *The Social System* (Glencoe, IL: The Free Press, 1951). For his criticism of the political right, see Parsons, "'McCarthyism' and American Social Tension: A Sociologist's View," *Yale Review* 44 (1955), 226–245. Howard Brick, *Transcending Capitalism: Visions of a New Society in Modern American Thought* (Ithaca, NY: Cornell Univ. Press, 2006), esp. "Talcott Parsons and the Evanescence of Capitalism," 121–151.
91. Talcott Parsons, "Weber's Methodology of Social Science," in Max Weber, *The Theory of Social and Economic Organization*, ed. Talcott Parsons (New York: The Free Press 1964/1947), 8–29; Parsons, *The Structure of Social Action.*
92. Parsons, "The Science Legislation and the Role of the Social Sciences," 660–662. Also see his other essays concerning this episode: "Science Legislation and the Social Sciences," *Political Science Quarterly* 62 (1947), 241–249; "National Science Legislation, Part 1: An Historical Review," *Bulletin of the Atomic Scientists* 2 (Nov. 6, 1946), 7–9; "National Science Legislation, Part 2: The Case for the Social Sciences," *Bulletin of the Atomic Scientists* 3 (Jan. 1947), 3–5.
93. Parsons, *The Social System*, 3. L. J. Henderson, *L. J. Henderson on the Social System*, ed. Bernard Barber (Chicago: Univ. of Chicago Press, 1970).

94. See materials in folder Cambridge Association of Natural and Social Scientists, 1946–1947, box 6, HUGFP 42.8.4, Parsons Papers, courtesy of the Harvard University Archives.
95. Parsons, "The Science Legislation and the Role of the Social Sciences," 666.
96. Parsons, ibid., 660; John H. Teeter to Talcott Parsons, Oct. 31, 1946, folder 1895, box 168, subseries 37, series 1, RG 2, SSRC Archives.
97. On the importance of Parsons's SSRC-commissioned essay for the discipline of sociology, see David P. Haney, *The Americanization of Social Science: Intellectuals and Public Responsibility in the Postwar United States* (Philadelphia: Temple Univ. Press, 2008), 22–45. As the reader will appreciate, however, I am not convinced by Haney's claim (p. 30) that in the postwar years sociologists such as Parsons, in a departure from the pre–World War II period, "did not focus upon the alleged similarities between the social and natural sciences or assert the predictability of their respective subject matter, and neither did they stress the difference between their work and moral commitment or policy advocacy, as had generations of earlier proponents of an objective scientific sociology."
98. Parsons, "A Basic National Resource," in Klausner and Lidz, *Nationalization of the Social Sciences*, scientific method at 42, essentially different at 43, total reconstruction at 111, pure research at 105.
99. Ibid., anti-longhair at 106, whole at 109.
100. Parsons, ibid., 107. On the lack of fit, see Minutes, Board of Directors Meeting, Sept. 13–16, 1948, pp. 11–14, folder 2100, box 358, series 9, RG 1, SSRC Archives.
101. S. S. Wilks to Pendleton Herring, Sept. 27, 1948, and Pendleton Herring to Dael Wolfle, no date. Both letters in folder 1896, box 168, subseries 37, series 1, RG 2, SSRC Archives. The SSRC project only came to a complete halt in 1951, after Parsons attempted a collaboration with another sociologist, John W. Riley, Jr. See Talcott Parsons to Pendleton Herring, Mar. 16, 1951, and Pendleton Herring to John W. Riley, Jr., Apr. 3, 1951. Both letters in folder SSRC Project, box 19, HUGFP 42.8.4, Parsons Papers, courtesy of the Harvard University Archives.
102. Gieryn, "The U.S. Congress Demarcates Natural Science and Social Science (Twice)," 74.
103. Robert K. Merton and Daniel Lerner, "Social Scientists and Research Policy," in Daniel Lerner and Harold D. Lasswell, eds., *The Policy Sciences: Recent Developments in Scope and Method* (Stanford, CA: Stanford Univ. Press, 1951), 282–307, quotation at 294.

CHAPTER 2. DEFENSE AND OFFENSE IN THE MILITARY SCIENCE ESTABLISHMENT

The first epigraph to chapter 2 is drawn from Don K. Price, *Government and Science: Their Dynamic Relation in American Democracy* (New York: New York Univ. Press, 1954), 96. Second epigraph: Charles Bray is quoted from page 3 of "The Technology of Human Behavior: Recommendations for Defense Support of Research in Psychology and the Social Sciences, A Report submitted to the Office of Science, Director of Defense Research and Engineering by the Research Group in Psychology and the Social Sciences, Smithsonian Institution, Washington, D.C., July 1960" (hereafter, 1960 Bray report), box 8, Record Unit 179, Smithsonian Institution, Research Group in Psychology and the Social Sciences Records, 1957–1963, Smithsonian Institution Archives, Washington, DC (hereafter, RU 179, Smithsonian Archives).

1. A. Hunter Dupree, *Science in the Federal Government: A History of Policies and Activities to 1940* (Cambridge, MA: Harvard Univ. Press, 1957), 375.
2. Charles G. Gant and Bertha Rubinstein, "Funds for Science: The Federal Government and Nonprofit Institutions," *Science* 117 (June 19, 1953), 669–676, dollar amounts on 669.
3. Daniel J. Kevles, "R&D and the Arms Race: An Analytic Look," in Everett Mendelsohn, Merritt R. Smith, and Peter Weingart, eds., *Science, Technology, and the Military*, vol. 2 (Boston: Kluwer, 1988), 465–480, percentages on 466.
4. Gant and Rubinstein, "Funds for Science," 670.
5. The following works provide useful discussions of wartime social science. On economics: Michael A. Bernstein, *A Perilous Progress: Economists and Public Purpose in Twentieth-Century America* (Princeton, NJ: Princeton Univ. Press, 2001); Philip Mirowski, *Machine Dreams: Economics Becomes a Cyborg Science* (New York: Cambridge Univ. Press, 2002). On psychology: James H. Capshew, *Psychologists on the March: Science, Practice, and Professional Identity in America, 1929–1969* (New York: Cambridge Univ. Press, 1999); Ellen Herman, *The Romance of American Psychology: Political Culture in the Age of Experts* (Berkeley: Univ. of California Press, 1995). On communications research: Christopher Simpson, *Science of Coercion: Communication Research and Psychological Warfare, 1945–1960* (New York: Oxford Univ. Press, 1994). On Soviet studies: David C. Engerman, *Know Your Enemy: The Rise and Fall of America's Soviet Experts* (New York: Oxford Univ. Press, 2009).
6. Barry M. Katz, *Foreign Intelligence: Research and Analysis in the Office of Strategic Services, 1942–1945* (Cambridge, MA: Harvard Univ. Press, 1989), xi.
7. Charles W. Bray, *Psychology and Military Proficiency: A History of the Applied Psychology Panel of the National Defense Research Committee* (Princeton, NJ: Princeton Univ. Press, 1948), xi, 229.
8. Jessica Wang, *American Science in an Age of Anxiety: Scientists, Anticommunism, and the Cold War* (Chapel Hill: Univ. of North Carolina Press, 1999), 283.
9. Mike F. Keen, *Stalking the Sociological Imagination: J. Edgar Hoover's FBI Surveillance of American Sociology* (Westport, CT: Greenwood Press, 1999), "No One above Suspicion: Talcott Parsons under Surveillance," 123–143; Susan Sperling, "Ashley's Ghost: McCarthyism, Science, and Human Nature," in Dustin M. Wax, ed., *Anthropology at the Dawn of the Cold War* (Ann Arbor, MI: Pluto Press, 2008), 17–36. Also David H. Price, *Threatening Anthropology: McCarthyism and the FBI's Surveillance of Activist Anthropologists* (Durham, NC: Duke Univ. Press, 2004).
10. On the Research and Analysis branch, see Katz, *Foreign Intelligence*. Robert D. Novak, "Subverting Bush at Langley," *Washington Post*, Dec. 24, 2007.
11. "Tongue-Tied," *Time*, Feb. 7, 1944. Allan M. Winkler, *The Politics of Propaganda: The Office of War Information, 1942–1945* (New Haven, CT: Yale Univ. Press, 1978).
12. David F. Krugler, *The Voice of America and the Domestic Propaganda Battles, 1945–1953* (Columbia: Univ. of Missouri Press, 2000), 1.
13. U.S. Congress, House, *Executive Sessions of the Senate Permanent Subcommittee on Investigations of the Committee on Government Operations 1953*, vol. 1 (Washington, DC: U.S. GPO, made publicly available in 2003). McCarthy chaired the committee and its subcommittee.
14. Paul Hoch, "The Crystallization of a Strategic Alliance: The American Physics Elite and the Military in the 1940s,'" in Mendelsohn et al., *Science, Technology and the Military*, 87–116, quotation at 87. Also see Gregg Herken, *Cardinal Choices: Presidential Science Advising from the Atomic Bomb to SDI* (New York: Oxford Univ. Press, 1992).

15. Price, *Government and Science*, 131.
16. Samuel A. Stouffer, "Sociology and the Strategy of Social Science," in Samuel A. Stouffer, *Social Research to Test Ideas: Selected Writings of Samuel A. Stouffer* (Glencoe, NY: Free Press, 1962), 1–10, quotations at 2.
17. Dorwin Cartwright, "Social Psychology in the United States during the Second World War," *Human Relations* 1 (1948), 333–352, quotation at 334; Fred Sheffield quoted on p. 209 in John A. Clausen, "Research on the American Soldier as a Career Contingency," *Social Psychology Quarterly* 47 (1984), 207–213.
18. The OSS's Research and Analysis Branch, where social scientists engaged in extensive discussions about the "epistemology of intelligence," provides an exception. See Katz, *Foreign Intelligence*.
19. Robert S. Lynd, "The Science of Inhuman Relations," *New Republic*, Aug. 29, 1949, 22–25, quotations at 22.
20. Nathan Glazer, "'The American Soldier' as Science: Can Sociology Fulfill Its Ambitions?" *Commentary*, Nov. 1949, 487–496, overpowering at 488, techniques at 489, understanding at 496.
21. Arthur Schlesinger, Jr., "The Statistical Soldier," *Partisan Review* 16 (1949), 852–856, whored at 852, neither at 855.
22. Alexander H. Leighton, *Human Relations in a Changing World: Observations on the Use of the Social Sciences* (New York: E. P. Dutton, 1949); Alexander H. Leighton, *The Governing of Men: General Principles and Recommendations Based on Experience at a Japanese Relocation Camp* (Princeton, NJ: Princeton Univ. Press, 1945); Daniel Lerner, *Sykewar: Psychological Warfare against Germany, D-Day to VE-Day* (New York: George W. Stewart, 1949); Karl M. Dallenbach, "The Emergency Committee in Psychology, National Research Council," *American Journal of Psychology* 59 (1946), 496–582; Bray, *Psychology and Military Proficiency*; Leonard W. Doob, "The Utilization of Social Scientists in the Overseas Branch of the Office of War Information," *American Political Science Review* 41 (1947), 649–667; Cartwright, "Social Psychology in the United States during the Second World War"; Paul F. Lazarsfeld, "The American Soldier—an Expository Review," *Public Opinion Quarterly* 13 (1949), 377–404.
23. Frederick Osborn, "The Social Sciences in the Service of Man," in *The Social Sciences at Mid-Century* (Minneapolis: Univ. of Minnesota Press, 1952), 3–11; Russell Sage Foundation, *Effective Use of Social Science Research in the Federal Services* (New York: Russell Sage Foundation, 1950).
24. Talcott Parsons, "Social Science: A Basic National Resource," in Samuel Z. Klausner and Victor M. Lidz, eds., *The Nationalization of the Social Sciences* (Philadelphia: Univ. of Pennsylvania Press, 1986), 79–101. Stuart Chase, *The Proper Study of Mankind*, 2d ed. (New York: Harper & Row, 1956/1948).
25. Stuart Chase, "Memorandum: Social Science Comes of Age," Jan. 1946, folder The Proper Study of Mankind, SS Outlines and Meetings, box 5, Stuart Chase Papers, Library of Congress, Washington, DC.
26. Carnegie Corporation, *1947 Annual Report*, 31.
27. Pendleton Herring to Dael Wolfle, no date, folder 1896, box 168, subseries 37, series 1, RG 2, Social Science Research Council Archives, Rockefeller Archive Center, Sleepy Hollow, NY.
28. Carnegie Corporation, *1949 Annual Report*, 24.
29. Lazarsfeld, "The American Soldier—an Expository Review," 377, 378.
30. Bray, *Psychology and Military Proficiency*, xi.

31. Chase, *The Proper Study of Mankind*, 41. For another example, see Leighton, *Human Relations in a Changing World*, 101–102.
32. Osborn, "The Social Sciences in the Service of Man," 4.
33. Parsons, "Social Science: A Basic National Resource," 95.
34. Russell Sage Foundation, *Effective Use of Social Science Research in the Federal Services*, 42.
35. Lazarsfeld, "The American Soldier—an Expository Review," 402.
36. Ron Robin, *The Making of the Cold War Enemy: Culture and Politics in the Military-Intellectual Complex* (Princeton, NJ: Princeton Univ. Press, 2001), 23.
37. James A. Smith, *The Idea Brokers: Think Tanks and the Rise of the New Policy Elite* (New York: The Free Press, 1991), 102.
38. William F. Ogburn, "Sociology and the Atom," *American Journal of Sociology* 51 (1946), 267–275, quotation at 274.
39. Lyle H. Lanier, "The Psychological and Social Sciences in the National Military Establishment," *American Psychologist* 4 (1949), 127–147, 1.5 percent on 131.
40. The RDB lasted until 1953, when the Defense Reorganization Act replaced it with the office of the new assistant secretaries for defense research and development and for engineering. Then in 1958 that office was abolished and its functions transferred to the new director of defense research and engineering. On the RDB, see Daniel J. Kevles, "K1 S2: Korea, Science, and the State," in Peter Galison and Bruce W. Hevly, eds., *Big Science: The Growth of Large-Scale Research* (Stanford, CA: Stanford Univ. Press, 1992), 312–333.
41. Lloyd V. Berkner, "Can the Social Sciences Be Made Exact?" *Proceedings of the IRE 48* (1960), 1376–1380, quotation at 1376.
42. Donald L. M. Blackmer, *The MIT Center for International Studies: The Founding Years, 1951–1969* (Cambridge, MA: MIT Center for International Studies, 2002).
43. Lanier, "The Psychological and Social Sciences in the National Military Establishment," 131–133.
44. Committee on Human Resources of the Joint Research and Development Board, Minutes of the First Meeting, July 1, 1947, Bush quotations at 9, 11, folder Human Resources Transcript of 1st Meeting, box 23, Record Group 330, Records of the Office of the Secretary of Defense, National Archives, Washington, DC (hereafter, RG 330, NA).
45. See "A Statement Concerning Research in Human Relations," May 7, 1948, folder 2, box 241, RG 330, NA.
46. Committee on Human Resources, "Research and Development in Human Resources in the National Military Establishment," June 28, 1948, pp. 51–52, quotations at 51, folder 1, box 245, RG 330, NA.
47. Committee on Human Resources, "Integrated Plan of Research and Development in Human Resources," Feb. 10, 1949, pp. vii, 3, 5, 9, folder 1, box 245, RG 330, NA.
48. Ibid., 12.
49. Budget figures from Kevles, "K1 S2," 319–320.
50. Memorandum, Dwight W. Chapman, Executive Director of the Committee on Human Resources, to the Chairman of the Research and Development Board, Jan. 24, 1951, folder 2, box 466, RG 330, NA.
51. Report of Working Group on Human Behavior under Conditions of Military Service, June 1951, box 411, RG 330, NA, limited at 396, old at 401.
52. Ibid., 402.
53. Memorandum, Aaron B. Nadel, Executive Director of the Committee on Human Resources, to Assistant Secretary of Defense, Research and Development, Dec. 21, 1953, folder 1, box 466, RG 330, NA.

54. The 24 percent figure comes from a 1966 National Science Foundation report and is noted on p. 1088 in Fred R. Harris, "Political Science and the Proposal for a National Social Science Foundation," *American Political Science Review* 61 (1967), 1088–1095. This 1966 report put the percentage of federal science funding going to the social sciences in the early 1950s at about 8 percent, compared with the 3 percent figure that is reported in the contemporary assessment I mention.
55. Gant and Rubinstein, "Funds for Science," 674.
56. Russell Sage Foundation, *Effective Use of Social Science Research in the Federal Services*, 19.
57. Lanier, "The Psychological and Social Sciences in the National Military Establishment," 136–140.
58. Ibid., 133–136.
59. W. L. Whitson, "The Growth of the Operations Research Office in the U.S. Army," *Operations Research* 8 (1960), 809–824.
60. On SORO, see Joy Rohde's essays: "Gray Matters: Social Scientists, Military Patronage, and Democracy in the Cold War," *Journal of American History* 96 (2009), 99–122; "From Expert Democracy to Beltway Banditry: How the Antiwar Movement Expanded the Military-Academic-Industrial Complex," 137–153, in Mark Solovey and Hamilton Cravens, eds., *Cold War Social Science: Knowledge Production, Liberal Democracy, and Human Nature* (New York: Palgrave Macmillan, 2012).
61. For a discussion of four HRRI projects carried out between 1949 and 1954, see George W. Croker, "Some Principles Regarding the Utilization of Social Science Research within the Military," in Charles Glock et al., *Case Studies in Bringing Behavioral Science into Use*, vol. 1 (Stanford, CA: Stanford Univ. Institute for Communication Research, 1961), 112–125. On the Harvard study, see David C. Engerman, "The Rise and Fall of Wartime Social Science: Harvard's Refugee Interview Project, 1950–1954," 25–43, in Solovey and Cravens, *Cold War Social Science*.
62. David Hounshell, "The Cold War, RAND, and the Generation of Knowledge, 1946–1962," *Historical Studies in the Physical and Biological Sciences* 27 (1997), 237–267, quotation at 265.
63. Williams quoted on pp. 61–62 in Bruce L. R. Smith, *The RAND Corporation: Case Study of a Nonprofit Advisory Corporation* (Cambridge, MA: Harvard Univ. Press, 1966).
64. "Conference of Social Scientists, September 14 to 19, 1947—New York," R-106 (Santa Monica, CA: RAND, June 9, 1948), http://www.rand.org/pubs/reports/R106, Goldhamer quotation in summary section "Outcome of the Conference," at viii.
65. John D. Marks, *The Search for the "Manchurian Candidate": The CIA and Mind Control* (New York: Times Books, 1979).
66. CIA Office of Research and Reports, "Soviet Defense Expenditures," Oct. 27, 1955, declassified, http://www.foia.cia.gov/docs/DOC_0000969867/DOC_0000969867.pdf.
67. X, "The Sources of Soviet Conduct," *Foreign Affairs* 25 (1947), 566–582, reprinted in George F. Kennan, *American Diplomacy, 1900–1950* (Chicago: Univ. of Chicago Press, 1951), 107–128, every at 118, among the peoples at 126–127.
68. On social science in U.S. information and propaganda programs, see Simpson, *Science of Coercion*; Scott Lucas, "Campaigns of Truth: The Psychological Strategy Board and American Ideology, 1951–1953," *International History Review* 18 (1996), 279–302; Timothy R. Glander, *Origins of Mass Communications Research during the American Cold War: Educational Effects and Contemporary Implications* (Mahwah, NJ: L. Erlbaum, 2000). After the Psychological Strategy Board was abolished in 1953, some of its responsibilities were transferred to the Operations Coordinating Board, also in the executive branch.

69. Christopher Simpson, "Universities, Empire, and the Production of Knowledge: An Introduction," xi–xxxiv, in Christopher Simpson, ed., *University and Empire: Money and Politics in the Social Sciences during the Cold War* (New York: The New Press, 1998), quotation at xii.
70. John G. Darley, "Five Years of Social Science Research: Retrospect and Prospect," in Harold S. Guetzkow, ed., *Groups, Leadership and Men: Research in Human Relations* (Pittsburgh: Carnegie Press, 1951), 3–15, quotations at 9–10.
71. Communications studies scholars have produced some excellent historical works on their field. I have found the following particularly useful: J. Michael Sproule, "Propaganda Studies in American Social Science: The Rise and Fall of the Critical Paradigm," *Quarterly Journal of Speech* 73 (1987), 60–78; Everett M. Rogers, *A History of Communication Study: A Biographical Approach* (New York: The Free Press, 1994); Simpson, *Science of Coercion*; Glander, *Origins of Mass Communications Research during the American Cold War*; Jefferson Pooley, "Fifteen Pages That Shook the Field: Personal Influence, Edward Shils, and the Remembered History of Mass Communication Research," *Annals of the American Academy of Political and Social Science* 608 (Nov. 2006), 130–156.
72. For a discussion of these developments by one of the main participants, see Leonard W. Doob, *Propaganda: Its Psychology and Technique* (New York: H. Holt, 1935), esp. 16–25.
73. Ibid., 5. Other IPA board members included the psychologist Hadley Cantril and the sociologist Alfred McClung Lee.
74. Wilbur Schramm, "Communications in Modern Society," in Wilbur Schramm, ed., *Communications in Modern Society: Fifteen Studies of the Mass Media* (Urbana: Univ. of Illinois Press, 1948), 1–6, quotations at 2.
75. Harold. D. Lasswell, "The Structure and Function of Communications in Society," in Wilbur Schramm and Donald F. Roberts, eds., *The Process and Effects of Mass Communication*, rev. ed. (Urbana: Univ. of Illinois Press, 1971/1954), 84–99, quotation at 88.
76. See Sproule, "Propaganda Studies in American Social Science"; Simpson, *Science of Coercion*; and Brett Gary, "Communication Research, the Rockefeller Foundation, and Mobilization for the War on Words, 1938–1944," *Journal of Communication* 46 (1996), 124–147.
77. Edward W. Barrett, *Truth Is Our Weapon* (New York: Funk & Wagnalls, 1953), totalitarian at 6–7, leaflets at 297.
78. Melvin L. DeFleur and Otto N. Larsen, *The Flow of Information: An Experiment in Mass Communication* (New York: Harper, 1958), effects at xv, 750,000 at 37, Birmingham study discussed on 47–48. Related studies targeted a wide variety of settings, including (xiv) "towns, grade schools, universities, colleges, urban centers, housing projects, boys camps, small groups, and communities struck by disaster."
79. Ibid., great deal at 1, Japan at 39.
80. Ibid., target at 38, weapon at 34, in motion at 33–34, actual at 38.
81. Ibid., 272.
82. Leonard W. Doob, *Public Opinion and Propaganda*, 2d ed. (Hamden, CT: Archon Books, 1966/1948), iii.
83. Simpson, *Science of Coercion*, 5.
84. Doob, *Public Opinion and Propaganda*, vi.
85. See Bernard Berelson, "The State of Communication Research," *Public Opinion Quarterly* 23 (1959), 1–6.
86. On the decision sciences, see Hunter Crowther-Heyck, *Herbert A. Simon: The Bounds of Reason in Modern America* (Baltimore: Johns Hopkins Univ. Press, 2005); Hunter

Heyck, "Producing Reason," 99–116, in Solovey and Cravens, *Cold War Social Science*; S. M. Amadae, *Rationalizing Capitalist Democracy: The Cold War Origins of Rational Choice Liberalism* (Chicago: Univ. of Chicago Press, 2003).

87. For a participant's view, see Florence N. Trefethen, "A History of Operations Research," in Joseph F. McCloskey and Florence N. Trefethen, eds., *Operations Research for Management*, vol. 1 (Baltimore: Johns Hopkins Univ. Press, 1954), 3–35. For historical accounts, see M. Fortun and S. S. Schweber, "Scientists and the Legacy of World War II: The Case of Operations Research (OR)," *Social Studies of Science* 23 (1993), 595–642; Philip Mirowski, "Cyborg Agonistes: Economics Meets Operations Research in Mid-Century," *Social Studies of Science* 29 (1999), 685–718.
88. King quoted in Philip M. Morse and George E. Kimball, *Methods of Operations Research* (New York: Wiley, 1951), 2.
89. Trefethen, "A History of Operations Research," 20–35.
90. Ibid., firmly at 3; Morse and Kimball, *Methods of Operations Research*, after from the preface.
91. Lawrence J. Henderson, "Organization for Operations Research," in McCloskey and Trefethen, *Operations Research for Management*, 69–79, quotations at 77, 78.
92. Philip M. Morse, "Progress in Operations Research," in McCloskey and Trefethen, *Operations Research for Management*, 99–116, Morse on economics at 116: Their mathematical tools included "queuing theory, linear programming, game theory, [and the] theory of optimum distribution of effort."
93. Stephen P. Waring, "Cold Calculus: The Cold War and Operations Research," *Radical History Review* 63 (1995), 29–51, quotation at 29–30; Ellis A. Johnson, "Introduction: The Executive, the Organization, and Operations Research," in McCloskey and Trefethen, *Operations Research for Management*, xi–xxiv, quotation at xxiii.
94. On the military's importance in the development of systems analysis and systems engineering, see Hounshell, "The Cold War, RAND, and the Generation of Knowledge, 1946–1962"; Thomas P. Hughes, *Rescuing Prometheus* (New York: Pantheon Books, 1998).
95. Albert J. Wohlstetter, "Economic and Strategic Considerations in Air Base Location," D-1114, Dec. 19, 1951, www.rand.org/about/history/Wohlstetter/D1114/D1114.html. For a discussion of Wohlstetter's analysis in the popular press, see Joseph Kraft, "RAND: Arsenal for Ideas," *Harper's Magazine* 221 (July 1960), 69–76, esp. 71–73.
96. John L. Kennedy, "The Uses and Limitations of Mathematical Models, Game Theory, and Systems Analysis in Planning and Problem Solution," in John C. Flanagan et al., *Current Trends: Psychology in the World Emergency* (Pittsburgh: Univ. of Pittsburgh Press, 1952), 97–116, quotation at 112.
97. Robert B. Miller, chair, Report of the Task Group on Design and Use of Man-Machine Systems, Nov. 15, 1959, quotations at 6, 9, box 7, RU 179, Smithsonian Archives.
98. John Von Neumann and Oskar Morgenstern, *Theory of Games and Economic Behavior*, 3d ed. (New York: John Wiley & Sons, 1964/1944), quotations at v, 1. On the history of game theory and the relevance of the national security context, see Philip Mirowski, "When Games Grow Deadly Serious: The Military Influence on the Evolution of Game Theory," in Craufurd D. W. Goodwin, ed., *Economics and National Security: A History of Their Interaction* (Durham, NC: Duke Univ. Press, 1991), 227–255; Robert Leonard, *Von Neumann, Morgenstern and the Creation of Game Theory: From Chess to Social Science, 1900–1960* (New York: Cambridge Univ. Press, 2010).

99. Von Neumann and Morgenstern, *Theory of Games and Economic Behavior*, quotations at ix, xxxi, 3–4. John D. Williams, *The Compleat Strategyst, Being a Primer on the Theory of Games of Strategy* (New York: McGraw-Hill, 1954), 8.
100. Martin Shubik, "Game Theory and the Study of Social Behavior: An Introductory Exposition," in Martin Shubik, ed., *Game Theory, and Related Approaches to Social Behavior: Selections* (New York: Wiley, 1964), 3–77, quotations at 8, 11.
101. Oskar Morgenstern, "The Theory of Games," *Scientific American* 189 (May 1949), 22–25, quotations at 25.
102. Shubik, "Game Theory and the Study of Social Behavior," 7.
103. On the history of deterrence theory, see Barry H. Steiner, *Bernard Brodie and the Foundations of American Nuclear Strategy* (Lawrence: Univ. of Kansas, 1991); Deborah W. Larson, "Deterrence Theory and the Cold War," *Radical History Review* 63 (1995), 86–109.
104. Robin, *The Making of the Cold War Enemy*, 8.
105. Bernard Brodie, "Strategy as a Science," *World Politics* 1 (1949), 467–488, Clemenceau quoted at 467.
106. Brodie, "Strategy as a Science," similarity at 476, genuine at 484, anti-theoretical at 486.
107. Larson, "Deterrence Theory and the Cold War."
108. Albert Wohlstetter, "The Delicate Balance of Terror," *Foreign Affairs* 33 (1959), 211–234, quotation at 211.
109. Ibid., quantitative at 213, extremely at 217, surprise at 231.
110. William W. Kaufmann, "The Requirements of Deterrence," in William W. Kaufmann, ed., *Military Policy and National Security* (Princeton, NJ: Princeton Univ. Press, 1956), 12–38, quotations at 25, 29.
111. David L. Snead, *The Gaither Committee, Eisenhower, and the Cold War* (Columbus: Ohio State Univ. Press, 1999); Amadae, *Rationalizing Capitalist Democracy*, 47.
112. John F. Kennedy, "Special Message to the Congress on the Defense Budget," Mar. 28, 1961, in *Public Papers of the Presidents: John F. Kennedy, 1961–1963* (Washington, DC: U.S. GPO, 1962–1964).
113. Harry Eckstein, ed., *Internal War: Problems and Approaches* (Westport, CT: Greenwood Press, 1980/1964).
114. Stephen Hosmer and S. O. Crane, eds., *Counterinsurgency: A Symposium* (Santa Monica, CA: RAND, 2006/1963), http://www.rand.org/pubs/reports/R412–1.html.
115. William A. Lybrand, ed., *Symposium Proceedings: The U.S. Army's Limited-War Mission and Social Science Research, March 26, 27, 28, 1962* (Washington DC: U.S. Army, SORO, June 1962), iii.
116. Mark Solovey, "Project Camelot and the 1960s Epistemological Revolution: Rethinking the Politics–Patronage–Social Science Nexus," *Social Studies of Science* 31 (2001), 171–206, esp. 180–182.
117. Gregory Palmer, *The McNamara Strategy and the Vietnam War: Program Budgeting in the Pentagon, 1960–1968* (Westport, CT: Greenwood Press, 1978).
118. Figures from U.S. Congress, House, Committee on Science and Astronautics, Subcommittee on Science, Research, and Development, *Technical Information for Congress*, report, serial A, ch. 6, "Congressional Response to Project Camelot," 91st Cong., 1st sess. (Washington, DC: U.S. GPO, Apr. 25, 1969, rev. May 15, 1971), 128–129.
119. Engerman, "The Rise and Fall of Wartime Social Science," Ferguson on 35–36, long hot at 36.

120. Memorandum, Aaron B. Nadel, Executive Director Committee on Human Resources, to the Assistant Secretary of Defense (R&D), July 20, 1953, folder 1, box 466, RG 330, NA.
121. James E. King, Jr., "Strategic Surrender: The Senate Debate and the Book," *World Politics* 11 (1959), 418–429, quotation from proposed amendment at 418, newspaper story at 419.
122. Paul Kecskemeti, *Strategic Surrender: The Politics of Victory and Defeat* (Stanford, CA: Stanford Univ. Press, 1958).
123. King, "Strategic Surrender," 422.
124. For a marvelous account of the 1960s that pays extensive attention to the New Left and critical intellectuals, see Howard Brick, *Age of Contradiction: American Thought and Culture in the 1960s* (New York: Twayne, 1998).
125. Irving L. Horowitz, "Arms, Policies and Games," *American Scholar* 31 (1961–1962), 94–107, quotation at 95.
126. Ibid., two players at 97, utilitarian at 98, mutual at 99, rules at 98.
127. Shubik, "Game Theory and the Study of Social Behavior," 10.
128. Horowitz, "Arms, Policies, and Games," scientific adequacy at 97, appearance at 101, far and dangers at 100.
129. Irving L. Horowitz, ed., *The Rise and Fall of Project Camelot: Studies in the Relationship between Social Science and Practical Politics* (Cambridge, MA: MIT Press, 1967).
130. Committee on Human Resources, "Integrated Plan of Research and Development in Human Resources," 19.
131. Dwight D. Eisenhower, "Farewell Radio and Television Address to the American People," in *Public Papers of the Presidents of the United States, 1960–61* (Washington, DC: U.S. GPO, 1961), 1035–1040, quotation at 1039.
132. Wilbur Schramm, chair, *Report of the Task Group on Basic Research in Persuasion and Motivation*, Feb. 18, 1960, p. 1, box 8, RU 179, Smithsonian Archives.
133. John G. Darley, "Psychology and the Office of Naval Research: A Decade of Development," *American Psychologist* 12 (1957), 305–323, quotations at 322.
134. "RAND Social Science Studies," July 20, 1948, 2, folder 11, box 30, William F. Ogburn Papers, Special Collections Research Center, University of Chicago Library, Chicago, IL. Nathan C. Leites, *The Operational Code of the Politburo* (New York: McGraw-Hill, 1951), xi.
135. Ralph L. Beals, *Politics of Social Research: An Inquiry into the Ethics and Responsibilities of Social Scientists* (Chicago: Aldine, 1969), 101.
136. "Text of Address by Nixon in San Francisco Assessing Challenge of Soviet Satellite," *New York Times*, Oct. 16, 1957.
137. James G. Miller, chair, "National Support for Behavioral Science," *Behavioral Science* 3 (1958), 217–227 (hereafter, 1958 Miller Report).
138. Ibid., Communist dogma at 218, other quotations at 219. One wonders if Nixon remained involved with this project, as the final report's unmistakable scientism seems at odds with his initial concerns.
139. For the complete list of consultants, see appendix I of the research group's "Final Summary Technical Report on Contract No. Nonrl1354," box 8, RU 179, Smithsonian Archives.
140. Charles W. Bray, "Toward a Technology of Human Behavior for Defense Use," *American Psychologist* 17 (1962), 527–541, key at 528, depend at 533, scientific theory at 528–529.
141. Ibid., 541.

142. Miller's recollections as presented in Otto N. Larsen, *Milestones and Millstones: Social Science at the National Science Foundation, 1945–1991* (New Brunswick, NJ: Transaction, 1992), 57, note 28.
143. 1958 Miller Report, recommendations on 218.
144. Again, see Miller's recollections as presented in Larsen, *Milestones and Millstones*, 57, note 28.
145. See note 7 of this chapter.
146. Charles Bray, "Notes on Planning Psychological and Social Science Research for DDRE, 1957, 1962," draft, Jan. 5, 1962, box 8, RU 179, Smithsonian Archives.
147. 1960 Bray report, 37.
148. Charles Bray to Ithiel de Sola Pool, Oct. 14, 1960, box 3, RU 179, Smithsonian Archives.
149. Bray, "Toward a Technology of Human Behavior for Defense Use," 541.

CHAPTER 3. VISION, ANALYSIS, OR SUBVERSION?

The epigraphs to chapter 3 are drawn from Bernard Berelson, "Introduction to the Behavioral Sciences," in Bernard Berelson, ed., *The Behavioral Sciences Today* (New York: Harper & Row, 1964), 1–11, quotation at 8; U.S. Congress, House, *Report of the Special Committee to Investigate Tax-Exempt Foundations and Comparable Organizations* (hereafter, Reece Committee *Report*), 83d Cong., 2d sess. (Washington, DC: U.S. GPO, 1954), 18.

1. Edwin R. Embree, "Timid Billions—Are the Foundations Doing Their Job?" *Harper's Magazine*, Mar. 1949, reprinted in U.S. Congress, House, Select Committee to Investigate Tax-Exempt Foundations and Comparable Organizations, *Tax-Exempt Foundations, Hearings* (hereafter, Cox Committee *Hearings*), 82d Cong., 2d sess. (Washington, DC: U.S. GPO, 1953), 299–308, quotation at 306.
2. Harry Alpert, with the assistance of Bertha W. Rubinstein, "The Role of the Foundation with Respect to Social Science Research," Apr. 15, 1954, percentages on pp. 10, 54, folder The Role of the Foundation with Respect to Social Science Research, Alpert-Rubinstein, 1954, NSF Historian's File (subsequently transferred to RG 307 of the National Archives: see chapter 1, note 10).
3. Raymond B. Fosdick, *The Story of the Rockefeller Foundation* (New York: Harper, 1952), v.
4. Ibid., 224.
5. Hunter Crowther-Heyck, "Patrons of the Revolution: Ideals and Institutions in Postwar Behavioral Science," *Isis* 97 (2006), 420–446, quotation at 437.
6. Roger L. Geiger, *Research and Relevant Knowledge: American Research Universities since World War II* (New York: Oxford Univ. Press, 1993), 100–101. Dollar amounts from "The Ford Foundation, Grants for Behavioral Sciences, 1952 through 1962," Aug. 1, 1963, folder 2, Berelson Papers, Ford Foundation Archives, Rockefeller Archive Center (hereafter, RAC), Sleepy Hollow, NY.
7. Ellen C. Lagemann, *The Politics of Knowledge: The Carnegie Corporation, Philanthropy, and Public Policy* (Middletown, CT: Wesleyan Univ. Press, 1989); Donald Fisher, *Fundamental Development of the Social Sciences: Rockefeller Philanthropy and the United States Social Science Research Council* (Ann Arbor: Univ. of Michigan Press, 1993).
8. Roger Geiger's standard account of philanthropic social science funding and Hunter Crowther-Heyck's important work say little about these challenges to scientism: Geiger, *Research and Relevant Knowledge*, 92–116; Crowther-Heyck, "Patrons of the Revolution." However, Alice O'Connor's work considers them: *Social Science for What?*

Philanthropy and the Social Question in a World Turned Rightside Up (New York: Russell Sage Foundation, 2007), esp. 48–70.

9. Howard M. Gitelman, *Legacy of the Ludlow Massacre: A Chapter in American Industrial Relations* (Philadelphia: Univ. of Pennsylvania Press, 1988).
10. U.S. Congress, Commission of Industrial Relations, *Final Report* (Washington, DC: U.S. GPO, 1916), menace at 83, recommendations on 220, 269.
11. "Memorial Policy in Social Science: Extracts from Various Memoranda and Dockets, October 1922 to July 1924," quotations from July 11, 1924, p. 21, folder 31, box 2, series 2, Laura Spelman Rockefeller Memorial Archives, RAC. Martin Bulmer and Joan Bulmer, "Philanthropy and Social Science in the 1920s: Beardsley Ruml and the Laura Spelman Rockefeller Memorial, 1922–29," *Minerva* 19 (1981), 347–407.
12. Carnegie Corporation, *1922 Annual Report*, 37. Also see Donald T. Critchlow, *The Brookings Institution, 1916–1952: Expertise and the Public Interest in a Democratic Society* (DeKalb: Northern Illinois Univ. Press, 1985).
13. Rockefeller Foundation, *1935 Annual Report*, 194, new program on 210–242.
14. Alvin Johnson, *Pioneer's Progress: An Autobiography* (New York: Viking, 1952), 332–348, quotations at 347. Peter Rutkoff and William Scott, *New School: A History of the New School for Social Research* (New York: Free Press, 1986), 84–106, 128–139.
15. Day quoted in John M. Jordan, *Machine-Age Ideology: Social Engineering and American Liberalism, 1911–1939* (Chapel Hill: Univ. of North Carolina Press, 1994), at 270. Frederick P. Keppel, *Philanthropy and Learning* (New York: Columbia Univ. Press, 1936), 22–23.
16. Inderjeet Parmar, "The Carnegie Corporation and the Mobilisation of Opinion in the United States' Rise to Globalism, 1939–45," *Minerva* 37 (1999), 355–378; Brett Gary, "Communication Research, the Rockefeller Foundation, and Mobilization for the War on Words, 1938–1944," *Journal of Communication* 46 (1996), 124–148.
17. Carnegie Corporation, *1948 Annual Report*, 23.
18. Charles Dollard, "Strategy for Advancing the Social Sciences," in G. Stanton Ford, ed., *The Social Sciences at Mid-Century* (Minneapolis: Univ. of Minnesota Press, 1952), 12–20, quotation at 17.
19. William E. Hocking to Chester I. Barnard, Sept. 6, 1946, p. 2, folder 162, box 21, series 900, RG 3.1, Rockefeller Foundation Archives (hereafter, RF Archives), RAC. Peter Novick, *That Noble Dream: The 'Objectivity Question' and the American Historical Profession* (New York: Cambridge Univ. Press, 1988), 288.
20. Chester I. Barnard to Dean Rusk, June 11, 1952, folder 304, box 56, series 900, RG 3.2, RF Archives, RAC. Lori Verstegen Ryan and William G. Scott, "Ethics and Organizational Reflection: The Rockefeller Foundation and Postwar 'Moral Deficits,' 1942–1954," *Academy of Management Review* 20 (1995), 438–461.
21. Dean Rusk to the Trustees, Mar. 27, 1953, folder 304, box 56, series 900, RG 3.2, RF Archives, RAC; Ryan and Scott, "Ethics and Organizational Reflection."
22. Nicolas Guilhot, ed., *The Invention of International Relations Theory: Realism, the Rockefeller Foundation and the 1954 Conference on Theory* (New York: Columbia Univ. Press, 2011).
23. Joseph H. Willits, "The Program in the Social Sciences," in *Rockefeller Foundation Trustees Bulletin* 1950, Confidential Monthly Report, June 1, 1950; Joseph Willits to Dean Rusk, Aug. 4, 1953, folder 304, box 56, series 900, RG 3.2. Both documents in RF Archives, RAC.
24. Carnegie Corporation, *1948 Annual Report*, 19.
25. Leffingwell in Cox Committee *Hearings*, 375–376. Reminiscences of Charles Dollard, 1969, Carnegie Corporation Project, p. 10, in the Columbia Center for Oral History,

Columbia University Libraries, New York, NY. Ad hoc and opportunistic in Reminiscences of Donald Young, 1967, Carnegie Corporation Project, p, 160, in ibid.

26. Roger L. Geiger, "American Foundations and Academic Social Science, 1945–1960," *Minerva* 26 (1988), 315–341, scaled back commitments on 325.
27. Ron Robin, *The Making of the Cold War Enemy: Culture and Politics in the Military-Intellectual Complex* (Princeton, NJ: Princeton Univ. Press, 2001); Rebecca Lowen, *Creating the Cold War University: The Transformation of Stanford* (Berkeley: Univ. of California Press, 1997), 194–202; David Engerman, *Know Your Enemy: The Rise and Fall of America's Soviet Experts* (New York: Oxford Univ. Press, 2009).
28. Francis X. Sutton, "The Ford Foundation: The Early Years," *Daedalus* 116 (1987), 41–92, financial assets on 52.
29. On Gaither, Ford, and RAND, see S. M. Amadae, *Rationalizing Capitalist Democracy: The Cold War Origins of Rational Choice Liberalism* (Chicago: Univ. of Chicago Press, 2003), 34–39.
30. H. Rowan Gaither, chairman, *Report of the Study for the Ford Foundation on Policy and Program* (Detroit: Ford Foundation, Nov. 1949), hereafter, 1949 *Study Report*. Also see "Report of the Trustees of the Ford Foundation, Sept. 27, 1950," reprinted in Cox Committee *Hearings*, 205–214.
31. 1949 *Study Report*, free peoples at 26–27, critical at 14, greatest at 98.
32. James G. Miller, "Toward a General Theory for the Behavioral Sciences," in Leonard D. White, *The State of the Social Sciences* (Chicago: Chicago Univ. Press, 1956), 29–65. Peter R. Senn, "What Is 'Behavioral Science?'—Notes toward a History," *Journal of the History of the Behavioral Sciences* 2 (1966), 107–122.
33. 1949 *Study Report*, polemical at 95, established at 93.
34. Gaither quoted in Dwight Macdonald, *The Ford Foundation: The Men and the Millions* (New York: Reynal, 1956), 80.
35. Gaither quoted in ibid., at 80–81; Report #002072, "The Ford Foundation Behavioral Sciences Program: Proposed Plan for the Development of the Behavioral Sciences Program," Dec. 1951, pp. 28, 29, 30, Ford Foundation Archives. In 1956 the BSP awarded $20,000 to Leo Strauss, a noted Chicago political theorist and harsh critic of scientism. But this grant is the exception, and it came at the tail end of BSP's existence. See Ford Foundation, *1956 Annual Report*, 56.
36. 1949 *Study Report*, long-range at 98, no one at 93–94, major and other fields and deficits at 94, storehouse at 93, applied and additional at 97.
37. 1949 *Study Report*, 51.
38. Amadae, *Rationalizing Capitalist Democracy*, 39.
39. 1949 *Study Report*, 91.
40. Ibid., 91.
41. Ibid., 54.
42. Ibid., 114.
43. Hoffman quoted in Alan R. Raucher, *Paul G. Hoffman: Architect of Foreign Aid* (Lexington: Univ. Press of Kentucky, 1985), 85.
44. Hans Speier, *Social Order and the Risks of War: Papers in Political Sociology*, 2d ed. (Cambridge, MA: MIT Press, 1969/1952). On Speier, see Christopher Simpson, *Science of Coercion: Communication Research and Psychological Warfare, 1945–1960* (New York: Oxford Univ. Press, 1994), 44–46.
45. See Marquis's APA presidential address: "Research Planning at the Frontiers of Science," *American Psychologist* 3 (1948), 430–438.

46. "The Ford Foundation, Behavioral Sciences Program," reel 1322, grant #54–47, section 5, p. 1, Ford Foundation Archives.
47. Reminiscences of Bernard Berelson, 1967, Carnegie Corporation Project, in the Columbia Center for Oral History. On Berelson, see David L. Sills, "Bernard Berelson: Behavioral Scientist," *Journal of the History of the Behavioral Sciences* 17 (1981), 305–311; Timothy R. Glander, *Origins of Mass Communications Research during the American Cold War: Educational Effects and Contemporary Implications* (Mahwah, NJ: L. Erlbaum, 2000), 77–82.
48. "Excerpt from Early Memorandum by Bernard Berelson on Development of Program Five," Oct. 1951, 1–4, folder 8, Berelson Papers.
49. Over the years, Berelson had five assistants (Dave McClelland, Waldemar Nielsen, Robert Knapp, Frank Sutton, and Dick Sheldon) and two supervisors (H. Rowan Gaither until the fall of 1953, and then William McPeak). See Bernard Berelson, "Five Year Report on Program V," Aug. 1956, p. 1, folder 4, Berelson Papers.
50. Sutton, "The Ford Foundation: The Early Years," nonsense at 72; useless in Author's interview with F. X. Sutton, Sept. 2, 1992.
51. Bernard Berelson, Ford Foundation Oral History Transcript, recorded July 1972, interviewed by C. T. Morrissey and R. J. Grele, p. 9, Ford Foundation Archives.
52. Donald K. David, "Business Responsibilities in an Uncertain World," *Harvard Business Review* (May 1949), 1–8, democracy at 1, organized and human at 6–7.
53. Donald David, Ford Foundation Oral History Transcript, recorded Mar. 1972, interviewed by C. T. Morrissey and R. J. Grele, p. 22, Ford Foundation Archives.
54. Bernard Berelson, "The Ford Foundation Behavioral Sciences Program Final Report, 1951–1957," Sept. 1957, p. 3, Berelson Papers.
55. John Lankford, *Congress and the Foundations in the Twentieth Century* (River Falls: Wisconsin State Univ. Press, 1964).
56. Other sociologists with FBI files included George Lundberg, Robert Lynd, C. Wright Mills, William Ogburn, Paul Lazarsfeld, Alfred McClung Lee, and Gunnar Myrdal (sometimes considered a sociologist, though an economist by training), and some members of the Frankfurt school. See Mike F. Keen, *Stalking the Sociological Imagination: J. Edgar Hoover's FBI Surveillance of American Sociology* (Westport, CT: Greenwood Press, 1999); David H. Price, *Threatening Anthropology: McCarthyism and the FBI's Surveillance of Activist Anthropologists* (Durham, NC: Duke Univ. Press, 2004). Also Dustin M. Wax, ed., *Anthropology at the Dawn of the Cold War: The Influence of Foundations, McCarthyism and the CIA* (Ann Arbor, MI: Pluto Press, 2008).
57. Report #002072, 15.
58. E. E. Cox, "Investigation of Certain Educational and Philanthropic Foundations," *Congressional Record—House*, Aug. 1, 1951, pp. A4833–A4834.
59. Cox Committee *Hearings*, Leffingwell at 373, Hoffman at 238.
60. Cox Committee *Hearings*, Fosdick at 762. On the problems faced by Soviet studies scholars during the McCarthy Era, see Engerman, *Know Your Enemy*, 41–42, 58–59.
61. U.S. Congress, House, *Final Report of the Select Committee to Investigate Foundations and Other Organizations*, hereafter Cox Committee *Final Report* (Washington DC: U.S. GPO, 1953), controversial at 4, risk capital at 3.
62. U.S. Congress, House, Special Committee to Investigate Tax-Exempt Foundations and Comparable Organizations, *Tax-Exempt Foundations, Hearings* (hereafter, Reece Committee *Hearings*), 83d Cong., 2d sess. (Washington, DC: U.S. GPO, 1954).
63. Reece Committee *Report*, power and act at 17, interlock at 18, Reece's LSE studies on 620; and his teaching of economics on p. 10924 in Reece, "America's Crisis in Education," *Congressional Record—House*, June 11, 1958, pp. 10921–10925.

64. McNiece in Reece Committee *Hearings*, 642. Alice O'Connor has noted that "for all the hysterics in Reece's charges, they were grounded in a crude awareness that the cultural underpinnings of old-fashioned, individualistic, 'free enterprise' capitalism were under threat." Furthermore, the foundations "were somehow implicated in its demise." See her essay "The Politics of Rich and Rich: Postwar Investigations of Foundations and the Rise of the Philanthropic Right," 228–248, quotation at 230, in Nelson Lichtenstein, ed., *American Capitalism: Social Thought and Political Economy in the Twentieth Century* (Philadelphia: Univ. of Pennsylvania Press, 2006).
65. *The Dodd Report, to the Reece Committee on Foundations* (New Canaan, CT: The Long House, 1954), 12. At the hearings, Dodd appeared as the first witness and testified at length: Reece Committee *Hearings*, 5–94.
66. O'Connor, *Social Science for What?*, 93.
67. William F. Buckley, Jr., *God and Man at Yale: The Superstitions of "Academic Freedom"* (Chicago: Henry Regnery, 1951), lix-lx; Friedrich A. Hayek, *The Counter-Revolution of Science: Studies on the Abuse of Reason* (Glencoe, IL: Free Press, 1952), 16, 15, 53, 94; Richard M. Weaver, *Ideas Have Consequences* (Chicago: Univ. of Chicago Press, 1948), 10, 1. During these years, many other books also attacked scientism and often social engineering at the same time. Good examples include Reinhard Bendix, *Social Science and the Distrust of Reason* (Berkeley: Univ. of California Press, 1951); Hans J. Morgenthau, *Scientific Man vs. Power Politics* (Chicago: Univ. of Chicago Press, 1946); Joseph W. Krutch, *The Measure of Man: On Freedom, Human Values, Survival, and the Modern Temper* (New York: Grosset and Dunlap, 1954); Leo Strauss, *What Is Political Philosophy? and Other Studies* (Glencoe, IL: Free Press, 1959); Helmut Schoeck and James W. Wiggins, eds., *Scientism and Values* (Princeton, NJ: Van Nostrand, 1960). Also see the works by Pitirim Sorokin and Rene Wormser cited below.
68. Pitirim A. Sorokin, "Declaration of Independence of the Social Sciences," *Social Science* 16 (1941), 221–229, quotation at 94; Sorokin, *Fads and Foibles in Sociology and Related Sciences* (Chicago: H. Regnery, 1956), 297. On Sorokin, see Barry Johnston, *Pitirim A. Sorokin: An Intellectual Biography* (Lawrence: Univ. Press of Kansas, 1995). The Reece Committee *Report* included an excerpt of a letter from Sorokin, on p. 64.
69. Albert H. Hobbs, *Social Problems and Scientism* (Harrisburg, PA: Stackpole, 1953). Hobbs in Reece Committee *Hearings*, love at 172, his full testimony on 114–187. During the hearings, the political scientists Kenneth W. Colegrove from Northwestern University and David N. Rowe from Yale University also testified about the scientistic and leftist tendencies within the social sciences and the foundations.
70. H. Rowan Gaither, Jr., "The Ford Foundation, Statement to The Special Committee to Investigate Tax-Exempt Foundations," in Reece Committee *Hearings*, erroneous at 1017; Report #010826, "Supplement A to Statement of H. Rowan Gaither, Jr., to The Special Committee to Investigate Tax Exempt Foundations," July 16, 1954, entire at 13, Ford Foundation Archives.
71. Herring in Reece Committee *Hearings*, father at 804–805, American tendency at 802, congressional at 800, his full testimony on 794–865.
72. Herring in ibid., 801, 838.
73. Reece Committee *Report*, 127.
74. Rene A. Wormser, *Foundations, Their Power and Influence* (New York: Devin-Adair, 1958).
75. Report #010628, Waldemar A. Nielsen to H. Rowan Gaither, Jr., "Summary and Evaluation of Reece Episode," Sept. 8, 1954, Ford Foundation Archives. *Minority Report* in Reese Committee *Report*, 417–432, complete at 421; Waldemar A. Nielsen, "Some Notes

on Press Coverage of the Reece Investigation," gauntlet at 4, folder 43, box 4, Nielsen Papers, Ford Foundation Archives.

76. Amadae, *Rationalizing Capitalist Democracy*, 38.
77. Nielsen, "Some Notes on Press Coverage of the Reece Investigation," gauntlet at 4, neutralizing at 10–11.
78. Reece Committee *Report*, 207.
79. *A Report on the Behavioral Sciences at the University of Chicago* (University of Chicago, Behavioral Sciences Self-Study Committee, 1954), 6.
80. Philippe Fontaine and Roger Backhouse, eds., "The Unsocial Social Science? Economics and Neighboring Disciplines since 1945," *History of Political Economy* 42, *Annual Supplement* (2010).
81. Jefferson Pooley and Mark Solovey, "Marginal to the Revolution: The Curious Relationship between Economics and the Behavioral Sciences Movement in Mid-Twentieth-Century America," in ibid., 199–233.
82. Berelson, "The Ford Foundation Behavioral Sciences Program Final Report, 1951–1957," basic resources at 11.
83. Reminiscences of Bernard Berelson, at 108.
84. Report #003156, William McPeak, "Behavioral Sciences Program, 1951–1957: Report and Appraisal," Dec. 1961, figures on p. 7, Ford Foundation Archives.
85. On the BSP consultants, see Report #002750, "The Ford Foundation Behavioral Sciences Division Report," June 1953, pp. 51–53, 60–67, Ford Foundation Archives.
86. On these grants before the BSP was fully operational, see Report #010818, "A Program in Behavioral Sciences Research," Sept. 1951; Report #010834, Pendleton Herring to H. Rowan Gaither, "Memorandum on The 5.1 Program"; and Report #002918, W. Allen Wallis, "The 1953–54 Program of University Surveys of the Behavioral Sciences," Apr. 18, 1953, representative at 1. All three reports in Ford Foundation Archives. $1.5 million noted in Ford Foundation, *1957 Annual Report*, 36.
87. Ford Foundation, *1957 Annual Report*, 34.
88. Donald David, Ford Foundation Oral History Transcript, 22.
89. Elbridge Sibley, "The Problem of Social Science Personnel: Recruitment or Training?" *SSRC Items* 1 (1947), 1–7, quotation at 5. Also see Elbridge Sibley, *The Recruitment, Selection, and Training of Social Scientists* (New York: SSRC, 1948).
90. Report of the Planning Group, "The Center for Advanced Study in the Behavioral Sciences," June 1952, reel 1322, #54–47, section 5, multiplier effect at 6, Ford Foundation Archives.
91. Ralph W. Tyler to Bernard Berelson, Apr. 19, 1954, reel 1322, grant #54–47, section 4, Ford Foundation Archives.
92. Sterling's comment recalled by Donald David in his Ford Foundation Oral History Transcript, at 22. The sociologist's quip is mentioned in Gene M. Lyons, *The Uneasy Partnership: Social Science and the Federal Government in the Twentieth Century* (New York: Russell Sage Foundation, 1969), at 280.
93. On the center's appeal, see "Report on the Center for Advanced Study in the Behavioral Sciences," Apr., 11, 1956, reel 1322, grant #54–47, section 3, p. 11, Ford Foundation Archives; Frank Sutton to Files, Jan. 12, 1955, reel 1322, grant #54–47, section 4, Ford Foundation Archives.
94. Ablest in Docket Excerpt, Board Meeting, Sept. 27–28, 1957, "Behavioral Sciences, Terminal Program," reel 1322, grant #54–47, section 1, Ford Foundation Archives.
95. R. Duncan Luce and Howard Raiffa, *Games and Decisions: Introduction and Critical Review* (New York: Wiley, 1957), x–xi. Also see Angela M. O'Rand, "Mathematizing Social

Science in the 1950s: The Early Development and Diffusion of Game Theory," in E. Roy Weintraub, ed., *Toward a History of Game Theory* (Durham, NC: Duke Univ. Press, 1992), 177–204, esp. 186. Ludwig von Bertalanffy, *General System Theory: Foundations, Development, Applications* (New York: George Braziller, 1968). Stanford's psychologist Leon Festinger also wrote his classic work on cognitive dissonance while on a fellowship at the center: *A Theory of Cognitive Dissonance* (Evanston, IL: Row, Peterson, 1957).

96. Berelson, "Five Year Report on Program V," programmatic at 3–4; Berelson, "Behavioral Sciences Program," fundamental at 2.
97. Dollar amounts and percentage noted on p. 29 in Peter J. Seybold, "The Ford Foundation and Social Control," *Science for the People* (May/June 1982), 28–31.
98. Frederick Mosteller, "The Role of the Social Science Research Council in the Advance of Mathematics in the Social Sciences," *SSRC Items* 28 (1974), 17–24. The committee's recommendation is quoted in Elbridge Sibley, *Social Science Research Council: The First Fifty Years* (New York: Social Science Research Council, 1974), 22.
99. Berelson, "Introduction to the Behavioral Sciences," 3.
100. There is a large historical literature on behavioralism in political science. One might start with the recent essay by Robert Adcock, "Interpreting Behavioralism," in Robert Adcock, Mark Bevir, and Shannon C. Stimson, eds., *Modern Political Science: Anglo-American Exchanges since 1870* (Princeton, NJ: Princeton Univ. Press, 2007), 180–208. On the Ford Foundation's role, see the essays by Peter J. Seybold, who offers a neo-Marxist analysis: "The Ford Foundation and the Triumph of Behavioralism in American Political Science," in Robert F. Arnove, ed., *Philanthropy and Cultural Imperialism: The Foundations at Home and Abroad* (Boston: G. K. Hall, 1980), 269–303, and "The Ford Foundation and the Transformation of Political Science," in Michael Schwartz, ed., *The Structure of Power in America: The Corporate Elite as a Ruling Class* (New York: Holmes & Meier, 1987), 185–198. For an excellent analysis of how Ford funding together with university politics shaped the rise of behavioralism in Stanford's Political Science Department, see Lowen, *Creating the Cold War University*, 194–212. Emily Hauptmann provides the most recent analysis in her essay "The Ford Foundation and the Rise of Behavioralism in Political Science," *Journal of the History of the Behavioral Sciences* 41 (2012), 154–173.
101. Austin Ranney, "The Committee on Political Behavior, 1949–64, and the Committee on Governmental and Legal Processes, 1964–72," *SSRC Items* 28 (1974), 37–41; Lucian W. Pye and Kay K. Ryland, "Activities of the Committee on Comparative Politics, 1954–70," *SSRC Items* 25 (1971), 6–9.
102. Pendleton Herring, "Political Science in the Next Decade," *American Political Science Review* 39 (1945), 757–766, essentially at 757, equivalent at 766. V. O. Key, "The State of the Discipline," *American Political Science Review* 52 (1958), 961–971, modest at 965.
103. Charles E. Lindblom, "The Science of 'Muddling Through,'" *Public Administration Review* 19 (1959), 79–88, incremental at 84. Gabriel A. Almond, "Comparative Political Systems," *Journal of Politics* 18 (1956), 391–409, quotation at 398.
104. Robert K. Merton, "The Mosaic of the Behavioral Sciences," in Berelson, *The Behavioral Sciences Today*, 246–272, quotation at 268; Sills, "Bernard Berelson," 306.
105. Seybold, "The Ford Foundation and the Triumph of Behavioralism in American Political Science," 271.
106. Geiger, *Research and Relevant Knowledge*, 102. Geiger also claims (p. 104) that during BSP's last few years, "it had largely swung around to accepting the academic status quo and to strengthening the social sciences on their own terms." But as I see it, this conclusion obscures the continuing debate about the nature and value of the

behavioral or social sciences both inside and outside the foundation. In addition, as noted later in this chapter, during the late 1950s and 1960s Ford continued to promote a great deal of policy-relevant social research and thus did not focus exclusively on or contribute mainly to "intellectual development," though Geiger claims so. (p. 108).

107. William Buxton, *Talcott Parsons and the Capitalist Nation-State: Political Sociology as a Strategic Vocation* (Toronto: Univ. of Toronto Press, 1985), 167–181.
108. Donald Young to Bernard Berelson, July 30, 1957, folder 354, box 41, series 4, Russell Sage Foundation Archives, RAC. Also see Donald R. Young, "Behavioral Science Application in the Professions," in Berelson, *The Behavioral Sciences Today*, 222–233.
109. Report #01525, "Working Paper for Meeting of Advisory Group on Mental Health Program," Mar. 28–29, 1955, Ford Foundation Archives; School of Industrial Management grant in Ford Foundation, *1957 Annual Report*, 25; Stanford grants in Ford Foundation, *1955 Annual Report*, 35, 37, respectively; MIT grant in Ford Foundation, *1952 Annual Report*, 52; grant to Population Council in Ford Foundation, *1954 Annual Report*, 55. The Ford Foundation's support for MIT's center is discussed in Blackmer, *The MIT Center for International Studies*. For an excellent discussion of foundation support for population research, see John Sharpless, "Population Science, Private Foundations, and Development Aid: The Transformation of Demographic Knowledge in the United States, 1945–1965," in Frederick Cooper and Randall M. Packard, eds., *International Development and the Social Sciences: Essays on the History and Politics of Knowledge* (Los Angeles: Univ. of California Press, 1997), 176–200.
110. Report #010584, Youth Development Program, "Report on Juvenile Delinquency," 1963, Ford Foundation Archives.
111. Report #004736, The Special Committee, "Public Affairs Program: Evolution (1950–1961) and Statement of Current Objectives and Policies," Dec. 1961, Ford Foundation Archives. On the wider context, see James B. Gilbert, *A Cycle of Outrage: America's Reaction to the Juvenile Delinquent in the 1950s* (New York: Oxford Univ. Press, 1986).
112. Richard A. Cloward and Lloyd E. Ohlin, *Delinquency and Opportunity: A Theory of Delinquent Gangs* (Glencoe, IL: Free Press, 1960), 211.
113. Mobilization for Youth, *A Proposal for the Prevention and Control of Delinquency by Expanding Opportunities* (New York: 1961), broad at iv, purpose at viii.
114. Harry S. Truman, "Inaugural Address," Jan. 20, 1949.
115. 1949 *Study Report*, 26–27.
116. Dollar amounts in Robert A. McCaughey, *International Studies and Academic Enterprise: A Chapter in the Enclosure of American Learning* (New York: Columbia Univ. Press, 1984), 192, 195.
117. State Department study noted on p. 222 in Edward H. Berman, "The Foundations' Role in American Foreign Policy: The Case of Africa, post 1945," in Arnove, *Philanthropy and Cultural Imperialism*, 203–232. Ford also provided extensive funding for the Congress for Cultural Freedom, an influential organization dedicated to anticommunist, liberal internationalism through the sponsorship of cultural and intellectual projects in Europe and elsewhere. See Volker R. Berghahn, *America and the Intellectual Cold Wars in Europe: Shepard Stone between Philanthropy, Academy, and Diplomacy* (Princeton, NJ: Princeton Univ. Press, 2001).
118. There is a large and sophisticated historical literature on area studies, development studies, modernization theory, and related foreign policy initiatives. I have found the following especially helpful: Michael E. Latham, *Modernization as Ideology: American Social Science and "Nation Building" in the Kennedy Era* (Chapel Hill: Univ. of

North Carolina Press, 2000); David C. Engerman, Nils Gilman, Mark H. Haefele, and Michael E. Latham, eds., *Staging Growth: Modernization, Development, and the Global Cold War* (Amherst: Univ. of Massachusetts Press, 2003); and Nils Gilman, *Mandarins of the Future: Modernization Theory in Cold War America* (Baltimore: Johns Hopkins Univ. Press, 2003).

119. Max F. Millikan and W. W. Rostow, *A Proposal: Key to an Effective Foreign Policy* (New York: Harper & Bros., 1957), 5–6.
120. Walt W. Rostow, *The Stages of Economic Growth: A Non-Communist Manifesto* (New York: Cambridge Univ. Press, 1960).
121. Report #002845, "Review Paper: Public Affairs: Gray Areas Program," Sept. 1964, pp. 17–18, Ford Foundation Archives. George A. Brager and Francis P. Purcell, eds., *Community Action against Poverty: Readings from the Mobilization Experience* (New Haven, CT: College & Univ. Press, 1967).
122. Alice O'Connor, "The Ford Foundation and Philanthropic Activism in the 1960s," in Ellen C. Lagemann, ed., *Philanthropic Foundations: New Scholarship, New Possibilities* (Bloomington: Indiana Univ. Press, 1999), 169–194, quotation at 170.
123. Blackmer, *The MIT Center for International Studies.*
124. Latham, *Modernization as Ideology*, 5. David Milne, *America's Rasputin: Walt Rostow and the Vietnam War* (New York: Hill & Wang, 2008).
125. Author's interview with Sutton, Sept. 24, 1992.
126. In *The Making of the Cold War Enemy*, Ron Robin notes (pp. 35–36) that political pressures during the McCarthy Era, including the Cox and Reece investigations, contributed to BSP's demise. But Robin makes no mention of the 1955 investigation of the Wichita Jury Study. Neither does one find a discussion of this investigation in other studies where it could be relevant, such as Richard Magat, *The Ford Foundation at Work: Philanthropic Choices, Methods, and Styles* (New York: Plenum Press, 1979), and Geiger, *Research and Relevant Knowledge.*
127. Harry Kalven, Jr., "Jury Project Statement," Oct. 12, 1955, appraise at 6, folder Questions of Method, box 7, University of Chicago Law School, Jury Project Records Addenda, Special Collections Research Center, University of Chicago Library (hereafter, Jury Project Records Addenda). Harry Kalven, Jr., and Hans Zeisel, *The American Jury* (Boston: Little, Brown, 1966), field research at v.
128. Report #002117, "Research Proposal on the Behavioral Sciences and the Law," May 19, 1952, Ford Foundation Archives.
129. John H. Schlegel, *American Legal Realism and Empirical Social Science* (Chapel Hill: Univ. of North Carolina Press, 1995), discussion of the Chicago Law and Behavioral Sciences Program on 238–244.
130. "Minimum" in Law Offices of Fleeson, Gooing, Coulson, and Kitch to Judge Phillips, Nov. 23, 1953, folder Jury Project Controversy, box 25, Jury Project Records Addenda. Notwithstanding the uproar over the secret recordings, other lines of research on the American jury continued, leading to dozens of articles and books. See Kalven, Jr., and Zeisel, *The American Jury.*
131. U.S. Congress, Senate, Committee on the Judiciary, Subcommittee to Investigate the Administration of the Internal Security Act and Other Internal Security Laws, *Recording of Jury Deliberations, Hearings*, 84th Cong., 1st sess., Oct. 12–13, 1955 (Washington, DC: U.S. GPO, 1955), ACLU representative Irving Ferman quoted at 186, U.S. attorney general quoted at 2. In this publication (hereafter, *Recording of Jury Deliberations, Hearings*), appendix VI contains the Chicago grant applications to Ford.

132. James J. Kilpatrick, *The Southern Case for School Segregation* (New York: Crowell-Collier, 1962), jurisprudence at 105. Kenneth B. Clark, *Prejudice and Your Child*, rev. ed. (Boston: Beacon, 1963), with the text of the 1954 decision in appendix 2. For an eye-opening historical analysis, see John P. Jackson, Jr., "The Triumph of the Segregationists: A Historiographical Inquiry into Psychology and the Brown Litigation," *History of Psychology* 3 (2000), 239–261.
133. *Recording of Jury Deliberations, Hearings*, 46, 86.
134. Ibid., Kalven at 62, Eastland at 85.
135. Wormser, *Foundations*, 251.
136. Public Law 919, U.S. Statutes at Large, 84th Cong., 2d sess., 70 (1956), 935–936.
137. On the case of left-leaning researchers, see Bernard Berelson, Ford Foundation Oral History Transcript, 20; on desegregation research, see Berelson, "Five Year Report on Program V," 9.
138. Berelson, "Five Year Report on Program V," 9.
139. Report #002686, F. X. Sutton, "Memorandum on the Comparative Value of the Encyclopedia of the Social Sciences," 1961, Ford Foundation Archives.
140. David's position as recalled by Berelson in his Ford Foundation Oral History Transcript, 23–25; David, Ford Foundation Oral History Transcript, 4; Berelson, "The Ford Foundation Behavioral Sciences Program Final Report, 1951–1957," disappointing at 16. Alvin Johnson, Letter to the Editor, *New York Times*, Dec. 22, 1954. In the 1960s Crowell, Collier and MacMillan published a new, multivolume encyclopedia as a commercial venture. Following the Reece Committee inquiry, Ford also decided to stop funding social science research on the impact of television. See David E. Morrison, "Opportunity Structures and the Creation of Knowledge: Paul Lazarsfeld and the Politics of Research," in David W. Park and Jefferson Pooley, eds., *The History of Media and Communication Research: Contested Memories* (New York: Peter Lang, 2008), 179–204.
141. Heald remark reported in "Notes of a telephone conversation with Frank Stanton, January 11, 1957" Reel 1322, Grant #54–47, section 4. Berelson, Ford Foundation Oral History Transcript, 63.
142. Report #003327, Henry T. Heald, Speech at The Brookings Institution, "Knowledge in the Service of Mankind," Nov. 17, 1960, p. 6, Ford Foundation Archives.
143. Berelson, "The Ford Foundation Behavioral Sciences Program Final Report, 1951–1957," unwise at 17, hurtful at 20.
144. Ibid., 20–21. After leaving Ford, Berelson continued to promote the behavioral sciences, as professor of behavioral science at the University of Chicago's Business School, as director of Columbia University's Bureau of Applied Social Research, and then as director of the Communication Research Program at the Population Council.
145. Magat, *The Ford Foundation at Work*, 31.
146. O'Connor, *Social Science for What?* 93.
147. Berelson, Ford Foundation Oral History Transcript, 63. Wyzanski's view recalled by Sutton, in Author's interview with Sutton, Sept. 24, 1992.
148. "Editorial," *Human Organization* 19 (1960), 49–50, quotation at 50.

CHAPTER 4. CULTIVATING HARD-CORE SOCIAL RESEARCH AT THE NSF

The epigraphs to chapter 4 are drawn from NSB Members' Books, 55th Meeting, 1958, Tab A, p. 7, NSF Library, NSF headquarters, Arlington, VA (hereafter, NSF Library); Harry Alpert, "The Government's Growing Recognition of Social Science," *Annals of the American Academy of Political and Social Science* 327 (1960), 59–67, quotation at 64.

1. NSF, *1950–51 Annual Report*, vii; Vannevar Bush, *Pieces of the Action* (New York: William Morrow, 1970), 65.
2. For brief but useful discussions of NSF's social science efforts during the period covered in this chapter, see J. Merton England, *A Patron for Pure Science: The National Science Foundation's Formative Years, 1945–57* (Washington, DC: NSF, 1982); Otto N. Larsen, *Milestones and Millstones: Social Science at the National Science Foundation, 1945–1991* (New Brunswick, NJ: Transaction Publishers, 1992); Daniel L. Kleinman and Mark Solovey, "Hot Science/Cold War: The National Science Foundation after World War II," *Radical History Review* 63 (1995), 110–139; Desmond King, "The Politics of Social Research: Institutionalizing Public Funding Regimes in the United States and Britain," *British Journal of Political Science* 28 (1998), 415–444.
3. William B. Cannon to Frederick C. Schmidt, Jr., p. 3, folder BOB Archives, NSF Historian's File (hereafter, NSF HF), (subsequently transferred to RG 307 of the National Archives: see chapter 1, note 10).
4. Harry Alpert, *Emile Durkheim and His Sociology* (New York: Columbia Univ. Press, 1939), 15.
5. Jennifer Platt, "The United States Reception of Durkheim's 'The Rules of Sociological Method,'" *Sociological Perspectives* 38 (1995), 77–105.
6. On Alpert's intellectual and professional development in relation to his NSF policy work, see Mark Solovey and Jefferson D. Pooley, "The Price of Success: Sociologist Harry Alpert, the NSF's First Social Science Policy Architect," *Annals of Science* 68 (2011), 229–260, quotation at 231, discussion of Alpert's views on interpretation on 238, 253.
7. Discussed in ibid., 255.
8. Author's conversation with Geoff Alpert, Apr. 17, 2011.
9. See Solovey and Pooley, "The Price of Success." Material on the Maryland job and the Gallup job offer comes from the Author's conversation with Geoff Alpert, Apr. 17, 2011.
10. Basic information about NSF's organizational structure and personnel comes from the agency's published annual reports. Also see England, *A Patron for Pure Science*; Larsen, *Milestones and Millstones*; and Toby A. Appel, *Shaping Biology: The National Science Foundation and American Biological Research, 1945–1975* (Baltimore: Johns Hopkins Univ. Press, 2000).
11. On Middlebush, see Larsen, *Milestones and Millstones*, 37.
12. Alpert, "The Knowledge We Need Most," *Saturday Review*, Feb. 1, 1958, 36–38, quotation at 37.
13. Ibid., embroil at 37. On NSF's efforts to insulate science from external pressures, see Kleinman and Solovey, "Hot Science/Cold War."
14. For the general counsel's memorandum (Mar. 17, 1953), see Harry Alpert with the assistance of Bertha W. Rubinstein, "The Role of the Foundation with Respect to Social Science Research," Apr. 15, 1954, appendix A, folder The Role of the Foundation with Respect to Social Science Research, Alpert-Rubinstein, 1954, NSF HF.
15. For lists of those Alpert consulted with, see Memo from Harry Alpert to Director Waterman, July 1, 1953, "Progress Report No. 1," appendix, and Memo from Harry Alpert to Dr. Waterman, Nov. 1, 1953, "Progress Report No. 2," appendix. Both documents in folder Role of the National Science Foundation with Respect to Social Research (Alpert, Mar. 1954), NSF HF.
16. Henry W. Riecken, "Underdogging: The Early Career of the Social Sciences in the NSF," in Samuel Z. Klausner and Victor M. Lidz, eds., *The Nationalization of the Social Sciences* (Philadelphia: Univ. of Pennsylvania Press, 1986), 209–225, quotation at 215.

17. John T. Wilson, "Psychology and 'Behavioral Science,'" paper presented at the University of Pittsburgh Conference on Interrelationships between the Behavioral Sciences, Mar. 11–12, 1954, folder Speeches—John T. Wilson, NSF HF.
18. Chester I. Barnard, "Social Science: Illusion and Reality," *American Scholar* 2 (1952), 359–360, quotations at 359.
19. William Scott, *Chester I. Barnard and the Guardians of the Managerial State* (Lawrence: Univ. Press of Kansas, 1992), 83. Barnard's view is reported in Memo, Harry Alpert to Director Waterman, May 22, 1953, folder [On Social Science Program], box 20, Subject Files of Dr. Alan T. Waterman, Record Group 307, Records of the National Science Foundation, National Archives (hereafter, Waterman Files in RG 307).
20. Memo from Harry Alpert to Director Waterman, July 1, 1953, "Position Paper No. 1," p. 1, folder Role of the National Science Foundation with Respect to Social Research (Alpert, Mar. 1954), NSF HF.
21. Harry Alpert to Dr. Waterman, Nov. 1, 1953, "Position Paper No. 2," p. 3, folder Reports on Social Science Program, NSF HF.
22. Ibid., effective at 1, experimental at 4.
23. NSB Minutes, 28th Meeting, Aug. 13, 1954, p. 7, NSF Library.
24. NSF, *1955 Annual Report*, 61.
25. "Annual Report FY 1956, Division of Mathematical, Physical, and Engineering Sciences," and "Annual Report of the Division of Biological and Medical Sciences for Fiscal Year 1956," folder Annual Reports of Offices and Divisions 1956, box 1, Waterman Files in RG 307.
26. "Inventory of Natural Science–Social Science Interdisciplinary Activities of National Science Foundation (through Fiscal Year 1954)," folder [On Social Science Program], box 20, Waterman Files in RG 307.
27. Harry Alpert with the assistance of Bertha W. Rubinstein, Feb. 1, 1956, "Progress Report No. 5," 3–4, folder [Reports on Social Science Program], NSF HF.
28. Ibid., 3.
29. Henry W. Riecken, "The National Science Foundation and the Social Sciences," *SSRC Items* 37 (1983), "Underdogging," 39–42, quotations at 40.
30. Harry Alpert, "The National Science Foundation and Social Science Research," *American Sociological Review* 19 (1954), 208–211, quotations at 210.
31. Alpert, "The Knowledge We Need Most," 37.
32. Alpert, "The National Science Foundation and Social Science Research," 210.
33. Author's interview with Bertha Rubinstein, Feb. 4, 1992.
34. Reece in *Congressional Record—House*, June 16, 1955, p. 8467.
35. Harry Alpert to Dr. Waterman, July 1, 1955, "Progress Report No. 4," 17–18, folder [Reports on Social Science Program], NSF HF.
36. John H. Teeter to Alan T. Waterman, June 30, 1955, folder Bush Papers, NSF HF.
37. John H. Teeter to Alan T. Waterman, Oct. 17, 1955, and frown in John H. Teeter to Alan T. Waterman, Nov. 10, 1955, folder [On Social Science Program], box 20, Waterman Files in RG 307.
38. Daniel J. Kevles, *The Physicists: the History of a Scientific Community in Modern America* (New York: Knopf, 1977), 362.
39. Careful consideration from Alan T. Waterman to John H. Teeter, Oct. 28, 1955, and very little from Alan T. Waterman to John H. Teeter, Dec. 22, 1955. Both documents in folder [On Social Science Program], box 20, Waterman Files in RG 307.
40. Harold J. Laski, *The Dangers of Obedience and Other Essays* (New York: Harper, 1930), 175.

41. Although Alpert made this observation during his tenure at the NSF, he wrote it not as a spokesman for the agency but as the president of the American Association of Public Opinion Research: Alpert, "Public Opinion Research as Science," *Public Opinion Quarterly* 20 (1956), 493–500, quotation at 494.
42. Eisenhower quoted in NSF, *1958 Annual Report*, 10, 11.
43. Budgetary information from Kevles, *The Physicists*, 386.
44. Javits in *Congressional Record—Senate*, May 13, 1957, pp. 6755–6756. Morse in *Congressional Record—Senate*, June 3, 1957, pp. 8211–8212.
45. Hubert H. Humphrey, *The Education of a Public Man: My Life and Politics*, ed. Norman Sherman (Garden City, NY: Doubleday, 1976), esp. 59–62; Carl Solberg, *Hubert Humphrey: A Biography* (New York: Norton, 1984), esp. 66–87. Miller Report in *Congressional Record—Senate*, Apr. 23, 1958, pp. 6252ff.
46. Alpert, "Congressmen, Social Scientists, and Attitudes toward Federal Support of Social Science Research," *American Sociological Review* 23 (1958), 682–686. On Kefauver's committee, see James Gilbert, *A Cycle of Outrage: America's Reaction to the Juvenile Delinquent in the 1950s* (New York: Oxford Univ. Press, 1986).
47. "Excerpts from Juvenile Delinquency, Report of Senate Committee on the Judiciary's Subcommittee on Juvenile Delinquency," in *Congressional Record—Senate*, June 3, 1957, p. 8212.
48. In my interview with her, Rubinstein recalled that Alpert made a point of cultivating relations with friendly legislators.
49. "Excerpts from Juvenile Delinquency Report of Senate Committee on the Judiciary's Subcommittee on Juvenile Delinquency," 8212.
50. Gilbert, *Cycle of Outrage*, 160.
51. Alan T. Waterman to Clyde C. Hall, July 29, 1957, folder Social Science Research Program, box 40, Waterman Files in RG 307.
52. Alpert, "Congressmen, Social Scientists, and Attitudes toward Federal Support of Social Science Research," 683.
53. Harry Alpert to Director Waterman, June 14, 1957, folder Social Science Research Program, box 40, Waterman Files in RG 307.
54. John G. Darley, "'New Man' in Washington for the Social Scientists," *Saturday Review*, Jan. 7, 1961, 91–92.
55. Riecken, "Underdogging," 216.
56. Henry W. Riecken, "Social Change and Social Science," in James A. Shannon, ed., *Science and the Evolution of Public Policy* (New York: Rockefeller Univ. Press, 1973), 135–155, quotations at 140–141. Author's interview with Rubinstein, Feb. 4, 1992. When Alpert left the NSF, Rubinstein became Riecken's assistant.
57. "Research is Aided in Social Sciences," *New York Times*, Dec. 5, 1960.
58. Alan T. Waterman, "National Science Foundation Program in the Social Sciences," July 8, 1958, folder Social Science Research Program, box 40, Waterman Files in RG 307.
59. Cressman in *Congressional Record—Senate*, June 3, 1957, p. 8213.
60. Charles O. Porter and Robert J. Alexander, *The Struggle for Democracy in Latin America* (New York: Macmillan, 1961), 1, 2.
61. Charles O. Porter, "Social Sciences Ignored in National Science Foundation Board Nominations," *Congressional Record—House*, Aug. 19, 1958, pp. 18591–18594.
62. Ogburn in U.S. Congress, Senate, A Subcommittee of the Committee on Military Affairs, *Hearings on Science Legislation (S. 1297 and Related Bills)*, 79th Cong., 1st sess., pts. 1–5, p. 774 (Washington, DC: U.S. GPO, 1945).

63. Porter, "Social Sciences Ignored," 18591; John Lear, "Our Report Card," *Saturday Review*, Feb. 1, 1958, 42.
64. Porter, "Social Sciences Ignored," 18591, 18592.
65. Memo from Program Director Social Science Research Program to Director, Sept. 26, 1958, folder Social Science Research Program, box 40, Waterman Files in RG 307.
66. NSF, *1959 Annual Report*, 44.
67. Memo from Director Waterman to Members of the National Science Board, in NSB Members' Books, 61st Meeting, Aug. 12, 1959, Tab H, NSF Library.
68. "Report of the Social Sciences Committee at Its Meetings on December 1, 1958, and January 23, 1959," in appendix VII to Minutes, Executive Session, 58th NSB Meeting, Jan. 23, 1959, p. 3, NSF Library.
69. Waterman quoted in Memo from Assistant Director for Social Sciences to Director, "Annual Review of the Division of Social Sciences, FY 1961," July 15, 1961, folder Social Science Reports 1959–1961, NSF HF.
70. Percentage from Larsen, *Milestones and Millstones*, 47.
71. Ibid., 46.
72. Bertha Rubinstein, "Federal Support of Unclassified, Extramural Social Science Research: 1953–1955," box 20, Waterman Files in RG 307. 1960 figures from "Federally Sponsored Research in the Social and Behavioral Sciences, A Report to the Federal Council on Science and Technology," Nov. 15, 1960, p. 7, folder Division of Social Sciences, box 68, Waterman Files in RG 307.
73. Harry Alpert, "The Growth of Social Research in the United States," in Daniel Lerner, ed., *The Human Meaning of the Social Sciences* (Cleveland: World Publishing Co., 1959), 73–86, quotations at 76–77.
74. Harry Alpert to Director Waterman, June 14, 1957, folder Social Science Research Program, box 40, Waterman Files in RG 307.
75. "Federally Sponsored Research in the Social and Behavioral Sciences," 8.
76. Early anthropological grants are noted in Memo from Head of Office of Social Sciences to Director, "Annual Review of the Office of Social Sciences, FY 1960," July 15, 1960, classical archaeology at 2, folder Social Science Reports 1959–1961, NSF HF. Other NSF grant titles from NSB Executive Session Minutes, 50th Meeting, Dec. 2, 1957, appendix V, NSF Library.
77. Lewis R. Binford, "Archaeology as Anthropology," *American Antiquity* 28 (1962), 217–225, quotations at 217. Also see Alice B. Kehoe, *The Land of Prehistory: A Critical History of American Archaeology* (New York: Routledge, 1998), 115–132.
78. Harry Alpert, "Demographic Research and the National Science Foundation," *Social Forces* 36 (1957), 17–21, quotation at 19. Memo from Head of Office of Social Sciences to Director, "Annual Review of the Office of Social Sciences, FY 1960," 5.
79. Memo from Head of Office of Social Sciences to Director, "Annual Review of the Office of Social Sciences, FY 1960," 5–6, quotation at 6.
80. On the Yerkes Laboratory grant, see Appel, *Shaping Biology*, 87. Harry Alpert, "Social Science, Social Psychology, and the National Science Foundation," *American Psychologist* 12 (1957), 95–98.
81. James H. Capshew, *Psychologists on the March: Science, Practice, and Professional Identity in America, 1929–1969* (New York: Cambridge Univ. Press, 1999), 179.
82. Memo from Head of Office of Social Sciences to Director, "Annual Review of the Office of Social Sciences, FY 1960," 9–11, quotation at 10. Also see Daniel H. Newlon, "The Role of the NSF in the Spread of Economic Ideas," in David C. Colander and

A. W. Coats, eds., *The Spread of Economic Ideas* (New York: Cambridge Univ. Press, 1989), 195–228.

83. Memo from Head of Office of Social Sciences to Director, "Annual Review of the Office of Social Sciences, FY 1960," 10.
84. Margaret W. Rossiter, "The History and Philosophy of Science Program at the National Science Foundation," *ISIS* 75 (1984), 95–104.
85. Steven Toulmin, "From Form to Function: Philosophy and History of Science in the 1950s and Now," *Daedalus* 106 (1977), 143–162; Steven Shapin, "Discipline and Bounding: The History and Sociology of Science as Seen through the Externalism-Internalism Debate," *History of Science* 30 (1992), 333–369.
86. The best-known Soviet Marxist analysis of science had been presented by Boris Hessen: "The Social and Economic Roots of Newton's 'Principia,'" in Nicolai I. Bukharin, ed., *Science at the Cross Roads: Papers Presented to the International Congress of the History of Science and Technology*, 2d ed. (London: Frank Cass, 1971/1931), 151–192. On the anti-Marxist thrust in history, philosophy, and sociology of science during the early Cold War years, see Peter Novick, *That Noble Dream: The "Objectivity Question" and the American Historical Profession* (New York: Cambridge Univ. Press, 1988), 295–301.
87. Carl G. Hempel, *Aspects of Scientific Explanation, and Other Essays in the Philosophy of Science* (New York: The Free Press, 1965). Karl R. Popper, *The Open Society and Its Enemies*, rev. ed. (Princeton, NJ: Princeton Univ. Press, 1950), 4–5, and *The Logic of Scientific Discovery* (New York: Basic Books, 1959). Also see Malachi H. Hacohen, *Karl Popper—the Formative Years, 1902–1945: Politics and Philosophy in Interwar Vienna* (New York: Cambridge Univ. Press, 2000); George A. Reisch, *How the Cold War Transformed Philosophy of Science: To the Icy Slopes of Logic* (New York: Cambridge Univ. Press, 2005).
88. I. B. Cohen, *Franklin and Newton: An Inquiry into Speculative Newtonian Experimental Science and Franklin's Work in Electricity as an Example Thereof* (Philadelphia: American Philosophical Society, 1956), and *The Birth of a New Physics* (Garden City, NY: Anchor, 1960).
89. Robert K. Merton, *The Sociology of Science: Theoretical and Empirical Investigations*, ed. Norman W. Storer (Chicago: Univ. of Chicago Press, 1973). David Hollinger, "The Defense of Democracy and Robert K. Merton's Formulation of the Scientific Ethos," *Knowledge and Society* 4 (1983), 1–15.
90. Grant information in NSB Executive Session Minutes, 46th Meeting, May 20, 1957, NSF Library.
91. Conference on History of Quantification in the Sciences, *Quantification: A History of the Meaning of Measurement in the Natural and Social Sciences* (Indianapolis: Bobbs-Merrill, 1961).
92. Memo from Director Waterman to Members of the National Science Board, Aug. 29, 1952, 15th Meeting, July 29, 1952, #2A, NSB Members' Books, NSF Library. George T. Mazuzan, "'Good Science Gets Funded': The Historical Evolution of Grant Making at the National Science Foundation," *Science Communication* 14 (1992), 63–90. However, other criteria besides good science have also played a role in the NSF review process, as explained in Marc Rothenberg, "Making Judgments about Grant Proposals: A Brief History of the Merit Review Criteria at the National Science Foundation," *Technology and Innovation* 12 (2010), 189–195.
93. Roger L. Geiger, *Research and Relevant Knowledge: American Research Universities since World War II* (New York: Oxford Univ. Press, 1993), 159.
94. Memo from Director Waterman to Members of the National Science Board, NSB Members' Books, 61st Meeting, Aug. 12, 1959, #H, NSF Library.

95. Dael Wolfle, "Social Science," *Science* 132 (1960), 1795.
96. Pendleton Herring in NSB Executive Session Minutes, 66th Meeting, June 29, 1960, appendix V, NSF Library.
97. Theodore M. Hesburgh, with Jerry Reedy, *God, Country, Notre Dame* (New York: Doubleday, 1990), 54–75; Michael O'Brien, *Hesburgh: A Biography* (Washington, DC: Catholic Univ. of America Press, 1998), 39–67.
98. Dwight D. Eisenhower, "First Inaugural Address," Jan. 20, 1953, in Davis N. Lott, *The Presidents Speak: The Inaugural Addresses of the American Presidents, from Washington to Clinton* (New York: Henry Holt, 1994), 302, 303.
99. Hesburgh quoted in O'Brien, *Hesburgh*, 62, 63.
100. Joan D. Goldhamer, "General Eisenhower in Academe: A Clash of Perspectives and a Study Suppressed," *Journal of the History of the Behavioral Sciences* 33 (1997), 241–259.
101. C. Wright Mills, *The Power Elite* (New York: Oxford Univ. Press, 1956), 350.
102. Eisenhower, "Presidential Address on the Role of the Intellectual in Governmental Affairs," May 1954, in U.S. Congress, House, Committee on Government Operations, Research and Technical Programs Subcommittee, *The Use of Social Research in Federal Domestic Programs, A Staff Study, Part 1: Federally-Financed Social Research—Expenditures, Status, and Objectives*, 90th Cong. 1st sess. (Washington, DC: U.S. GPO, 1967), 163–177, atheistic at 164, truth at 167.
103. Hesburgh, *God, Country, Notre Dame*, 7.
104. Theodore M. Hesburgh, *God and the World of Man* (Notre Dame, IN: Univ. of Notre Dame Press, 1960), 5.
105. Alpert, "The Social Sciences: Problems, Issues and Suggested Resolutions," Apr. 15, 1958, in NSB Members' Books, 54th Meeting, June 24, 1958, Tab R, NSF Library.
106. Theodore M. Hesburgh, "Report of Committee on Social Sciences," May 18, 1958, in NSB Members' Books, 54th Meeting, June 24, 1958, Tab Q, NSF Library.
107. Hesburgh quoted in Riecken, "Underdogging," 215.
108. Hesburgh, "Report of Committee on Social Sciences."
109. Ibid.
110. Unidentified member and Stratton quoted in NSB Executive Session Minutes, 54th Meeting, June 28–30, 1958, pp. 6, 7, NSF Library.
111. Warren Weaver, *Scene of Change: A Lifetime in American Science* (New York: Scribner, 1970), 154.
112. NSB Executive Session Minutes, 54th Meeting, June 28–30, 1958, pp. 6–7.
113. McLaughlin's view noted on p. 7 in ibid.
114. Kevin McCann, *Man from Abilene* (Garden City, NY: Doubleday, 1952). McCann quoted in NSB Members' Books, 55th Meeting, 1958, Tab A, NSF Library, p. 7.
115. "Report of the Social Sciences Committee" in NSB Members' Books, 56th Meeting, 1958, pp. 3–4, NSF Library.
116. Riecken, "Underdogging," quotation at 217, Rubinstein at 217.
117. Ibid., 214.
118. Alpert, "The Knowledge We Need Most."
119. Ibid., 38.
120. Alpert, "The Government's Growing Recognition of Social Science," 61.
121. Alpert, "The Growth of Social Research in the United States," 82.
122. Ibid., note 2.
123. 1951 NSB Resolution included in Alpert, Position Paper No. 2.
124. Bruce L. R. Smith, *American Science Policy since World War II* (Washington, DC: Brookings Institution Press, 1990), 51.

125. NSF, *1959 Annual Report*, 11.
126. Discussed by Riecken in Memo from Assistant Director Social Sciences Program to Director, "Annual Review of the Division of Social Sciences, FY 1961," July 15, 1961, p. 8.
127. See Memorandum, Henry W. Riecken to Leland J. Haworth, Sept. 19, 1963, folder Political Science 1964, box 11, O/D, July 1963–Dec. 1964, RG 307, National Archives. All documents cited in the notes below come from this box, unless otherwise indicated.
128. Evron M. Kirkpatrick to Senator George S. McGovern, Nov. 26, 1963, and Evron M. Kirkpatrick to Mr. Haworth, Nov. 26, 1963. Both documents in folder Political Science 1963.
129. Bowen G. Dees to Evron M. Kirkpatrick, July 31, 1963, folder Political Science 1964.
130. Henry W. Riecken to Leland J. Haworth, Sept. 19, 1963, folder Political Science 1963.
131. 1963 divisional meeting minutes quoted in ibid.
132. Ibid.
133. Kirkpatrick to McGovern, Nov. 26, 1963.
134. "Statement of Evron M. Kirkpatrick, Executive Director, the American Political Science Association, Protesting the Discrimination Against Political Science by the National Science Foundation," Dec. 12, 1963, attached to Evron M. Kirkpatrick to Mr. Haworth, Jan. 2, 1964, folder Political Science 1964.
135. "NSF and Behavioral Science," *American Behavioral Scientist* 7 (Sept. 1963), 70.
136. Henry W. Riecken, "Background Paper prepared for Committee on Social Sciences," National Science Board, Nov. 17–18, 1960, p. 1, folder Office of Social Sciences, box 53, Waterman Files in RG 307.
137. T. M. Hesburgh to Malcolm M. Willey, Nov. 3, 1960, folder NSB Files, 1959–1965, NSF HF.

CONCLUSION

1. Irving L. Horowitz, ed., *The Rise and Fall of Project Camelot: Studies in the Relationship between Social Science and Practical Politics* (Cambridge, MA: MIT Press, 1967). For historical analysis, see Mark Solovey, "Project Camelot and the 1960s Epistemological Revolution: Rethinking the Politics–Patronage–Social Science Nexus," *Social Studies of Science* 31 (2001), 171–206; Joy Rohde, "Gray Matters: Social Scientists, Military Patronage, and Democracy in the Cold War," *Journal of American History* 96 (2009), 99–122.
2. Mark Solovey, "Senator Fred Harris's National Social Science Foundation Proposal: Reconsidering Federal Science Policy, Natural Science–Social Science Relations, and American Liberalism during the 1960s," *Isis* 103 (2012), 54–82; Thomas F. Gieryn, "The U.S. Congress Demarcates Natural Science and Social Science (Twice)," in his book *Cultural Boundaries of Science: Credibility on the Line* (Chicago: Univ. of Chicago Press, 1999), 65–114, esp. 84–93, 101–108.
3. J. David Hoeveler, *Watch on the Right: Conservative Intellectuals in the Reagan Era* (Madison: Univ. of Wisconsin Press, 1991); Godfrey Hodgson, *The World Turned Right Side Up: A History of the Conservative Ascendancy in America* (Boston: Houghton Mifflin, 1996).
4. According to Joy Rohde, though the antiwar movement helped to push military-funded institutes off university campuses, success in this respect also meant less academic oversight of military-funded social research. See her essay "From Expert Democracy to Beltway Banditry: How the Antiwar Movement Expanded the Military-Academic-Industrial Complex," in Mark Solovey and Hamilton Cravens, eds., *Cold War Social Science: Knowledge Production, Liberal Democracy, and Human Nature* (New York: Palgrave Macmillan, 2012), 137–153.

5. Andrew Rich, *Think Tanks, Public Policy, and the Politics of Expertise* (New York: Cambridge Univ. Press, 2004); Alice O'Connor, *Social Science for What? Philanthropy and the Social Question in a World Turned Rightside Up* (New York: Russell Sage Foundation, 2007). For sympathetic accounts of conservative foundations that blend historical analysis with positive assessments, see Lee Edwards, *The Power of Ideas: The Heritage Foundation at 25 Years* (Ottawa, IL: Jameson Books, 1997); John J. Miller, *A Gift of Freedom: How the John M. Olin Foundation Changed America* (San Francisco: Encounter Books, 2005).
6. Otto N. Larsen, *Milestones and Millstones: Social Science at the National Science Foundation, 1945–1991* (New Brunswick, NJ: Transaction Publishers, 1992), ch. 6.
7. Hugh Gusterson, "The U.S. Military's Quest to Weaponize Culture," *Bulletin of the Atomic Scientists*, Web edition, June 20, 2008, http://thebulletin.org/web-edition/columnists/hugh-gusterson/the-us-militarys-quest-to-weaponize-culture. Also see David H. Price, *Weaponizing Anthropology: Social Science in Service of the Militarized State* (Oakland, CA: AK Press, 2011).
8. "American Anthropological Association Executive Board Statement on the Human Terrain System Project," Oct. 31, 2007, http://www.aaanet.org/pdf/ed_resolution_110807.pdf.
9. Alfred W. McCoy, *A Question of Torture: CIA Interrogation, from the Cold War to the War on Terror* (New York: Metropolis Books, 2006).

INDEX

Page numbers in italics refer to figures.

ABOUT THE AUTHOR

MARK SOLOVEY teaches at the Institute for the History and Philosophy of Science and Technology at the University of Toronto. His research focuses on the political, institutional, and intellectual histories of the social sciences in the United States, especially since World War Two. He has several articles in scholarly journals, including *Annals of Science*, *History of Political Economy*, *Isis*, *Journal of the History of the Behavioral Sciences*, *Radical History Review*, and *Social Studies of Science*. He was the guest editor for a *Social Studies of Science* issue on science in the Cold War. He is also the coeditor of *Cold War Social Science: Knowledge Production, Liberal Democracy, and Human Nature* (2012).

CPSIA information can be obtained
at www.ICGtesting.com
Printed in the USA
LVHW042300171219
640837LV00006B/12/P